THE COSMIC FRONTIERS OF GENERAL RELATIVITY

Happy Reading

Grandpa

THE COSMIC FRONTIERS

OF GENERAL RELATIVITY

WILLIAM J. KAUFMANN, III

Astrophysics-Relativity Group
Department of Physics
San Diego State University
and
Department of Astronomy
University of California, Los Angeles

LITTLE, BROWN AND COMPANY
Boston   Toronto

Library of Congress Catalog Card No. 76-46800

Second Printing

Published simultaneously in Canada
by Little, Brown & Company (Canada) Limited

Printed in the United States of America

Editorial production supervision: Freda Alexander
Cover design: Richard S. Emery
Interior design: Judy Arisman
Manuscript editing: Marret McCorkle and Robin Watkins
Art: Vantage Art

Credits for Illustrations

The art, for the most part, has been redrawn from originals by Joy Wohl. Credits for photographs and adapted illustrations follow:

Cover photo: See description on back cover.

Chapter 1: *1-1* E. C. Krupp, Griffith Observatory. *1-4* Hale Observatories.

Chapter 4: *4-4, 4-6, 4-8,* and *4-15* Yerkes Observatory. *4-12* and *4-13* Lick Observatory.

Chapter 5: *5-1* Lick Observatory. *5-2* NASA, Jet Propulsion Laboratory. *5-4* NASA. *5-5, 5-7,* and *5-8* Hale Observatories. *5-10* and *5-11* Adapted from *Gravitation* by Charles W. Misner, Kip S. Thorne, and John Archibald Wheeler. W. H. Freeman and Company. Copyright © 1973.

Chapter 6: *6-2* and *6-10* Hale Observatories. *6-6, 6-7, 6-8,* and *6-9* Lick Observatory.

Chapter 7: *7-1* NASA, Jet Propulsion Laboratory. *7-2* Adapted from Lund Observatory; Griffith Observatory. *7-4* Lunar and Planetary Laboratory photograph. *7-5* and *7-9* Lick Observatory.

Chapter 10: *10-1* Adapted with permission of John Archibald Wheeler, and from R. Ruffini and J. A. Wheeler, "Relativistic Cosmology and Space Platforms" in *Proceedings of the Conference on Space Physics,* published by European Space Research Organization, Paris, France, fig. 14, 1971.

To Lee with love

PREFACE

A dramatic resurgence of interest in the general theory of relativity has been one of the most remarkable developments in astronomy during the past ten years. The discovery of pulsars in 1967 inspired astrophysicists to make detailed calculations of the structure of neutron stars; these results show that *no* stable models exist for stars with masses greater than three solar masses. Prompted by this and similarly exciting discoveries, theoretical astrophysicists attacked the problem of black holes with unprecedented vigor. Their efforts have brought about an extraordinarily wide range of discoveries.

Exciting as these discoveries are, though, they have been available only to professional astronomers and physicists, who have the mathematical background necessary to read the professional literature. The interested layperson and the curious student have virtually nowhere to turn to for detailed information on topics like black holes, Cygnus X-1, or warped space-time.

This book was written in response to this need. It is nontechnical in its approach, and requires that the reader have only an inquiring mind. Little mathematical or astronomical background is assumed or required, and much attention is given to expressing concepts in language that a nonspecialist can readily understand. In addition, graphs and diagrams are used throughout to present clearly many concepts in general relativity that might otherwise require complex mathematics. The book contains over 200 illustrations, among them embedding diagrams, Penrose diagrams, and views from the windows of spacecrafts falling through wormholes.

Because it can provide a stimulating new dimension to a first encounter with astronomy, this book can serve as a supplement in an introductory astronomy course, or as a more central text in second-level courses frequently offered to liberal arts students interested in pursuing some topics in astronomy in greater detail. The book includes all the relevant concepts from a general astronomy course, and great care has been taken to make the treatment lively and interesting for those in and outside the classroom.

In the fall of 1974, I went on sabbatical leave from the directorship of the Griffith Observatory and returned to the Astrophysics-Relativity Group at the California Institute of Technology. I am grateful to members of the Physics Department and especially to Kip S. Thorne, for the many courtesies shown me. Primarily as a result of my being at Caltech, this book contains discoveries that have not yet been transmitted to the professional journals. In this regard, I am deeply grateful to C. T. Cunningham and S. W. Hawking for numerous enlightening discussions and for access to their calculations prior to publication.

I wish to thank G. O. Abell at the University of California, Los Angeles, and D. M. Eardley at Yale University for their suggestions and comments on the manuscript. I am also grateful to the Astronomy Department at UCLA for inviting me to teach the selected-topics-for-nonmajors course in the fall of 1975 and in the summer of 1976, as this gave me the opportunity to test this manuscript with students.

The originals of most of the drawings were prepared by Joy Wohl, whose talents and patience deserve the highest credit. Finally, I am deeply grateful to Louise Nelson for typing the manuscript.

CONTENTS

THE COSMIC FRONTIERS OF GENERAL RELATIVITY

1 AN ORIENTATION IN SPACE-TIME

For thousands of years people have gazed into the star-filled nighttime sky and felt a sense of mystery and wonder. Even in the most ancient times, before recorded history, people have marveled at the orderly workings of the heavens. The rising and setting of the sun, the silver moon going through its phases, the drama of an eclipse, and the wanderings of the planets among the constellations of the zodiac proved sufficient to inspire our ancestors to take up the study of astronomy.

In retrospect, the depth of insight and the dedication to knowledge demonstrated in many of these ancient civilizations is truly incredible. The architects of the pyramids and the builders of Stonehenge clearly had a great wealth of information at their fingertips, astronomical information that could only have been gathered over decades or generations of patient, careful observation. From the extensiveness of these observations it can only be

1

FIGURE 1-1. *Stonehenge.*
This astronomical monument,
built almost 5,000 years ago,
stands in mute testimony to
the insight and ingenuity of
early people. (Griffith Observatory)

concluded that ancient men and women were far more interested in astronomy than people today living in the so-called "space age." Indeed, in a certain sense it may be said that the building of Stonehenge 5,000 years ago was a far more impressive accomplishment for humanity than traveling to the moon during the last decade (Figure 1-1). This is especially apparent when we consider the minuscule cost of our space program and how it has touched our lives. At the height of the Apollo program, the annual cost was comparable to the amount of money Americans spend each year on dog food, one-third of what they spend on cigarettes, and one-seventh of what they spend on liquor. Compare this with the fact that thousands of years ago urban centers, such as those found in Central America, were built with key astronomical orientations in mind. When we stop and consider the phenomenal accomplishments of early civilizations in astronomy and how their astronomical knowledge permeated everything from the construction of monuments and temples to the aligning of city streets, scientific efforts at Palomar or in Skylab take on a new perspective. It is almost as if we actually *see* the cosmos from a viewpoint fundamentally different from our predecessors'.

Regardless of one's viewpoint or orientation, all knowledge of the

2

universe must begin with some sort of observations. But few human beings are satisfied with endless observations alone. It is not enough to go out night after night and merely record the positions of the stars and planets or to take endless photographs. Instead, at some point we are all inclined to ask why things are the way they are. At some point we want to know why and how the planets move, or we want to know why galaxies have certain sizes and shapes. To various degrees, over the ages human beings have always felt that there was some level of order to the universe. The regular rising and setting of the sun over the years, or the moon going through the same phases every four weeks, implies order rather than chaos. It is this appearance of order that gives rise to the hope of finding a deeper understanding of the universe.

A collection of ideas or a hypothesis that expresses this understanding of the universe and from which astronomical observations can be explained is called a *cosmology.* Every civilization and every religion to appear on our planet has had a cosmology at the core of its teachings. The nature and content of a cosmology is profoundly dependent on the culture from which the cosmology originated. In the most ancient civilizations, cosmological ideas were said to be transmitted from the gods through high priests and occult initiates. In later times, beginning with the Greeks, people relied more on direct observations of the heavens. In view of the wide range of accepted methods for obtaining an understanding of the universe, we are not surprised by the colorful and often contradictory theories of the cosmos which were prevalent in different societies. Some of these theories consist of myths and legends that reveal far more about the psychology of a civilization than the physical nature of the universe; other ideas demonstrate a rudimentary application of what we today might call the "scientific method." And we wonder if it is possible that thousands of years from now the astronomy and astrophysics of the twentieth century will be looked upon as fantasy and myth.

Of course, a cosmology is dependent on available observations, and no intelligent astronomer would deny that theories should be modified or even abandoned as new information comes to light. However, in a much more profound sense, a cosmology is affected by a wide range of axioms and assumptions that frequently are unexamined and unquestioned. For example, modern psychology teaches that the way in which children learn to process sensory information and formulate concepts and ideas results in a particular way of seeing the world. Thus human beings systematically exclude or include certain data according to prior psychological conditioning. And modern scientists, who with wide-eyed innocence protest their objectivity, are actually as biased as the Catholic priests who persecuted Galileo. The priests who declined Galileo's invitation and *refused* to look through his telescope at the moon, Venus, or Jupiter, did so simply because this was not an accepted methodology for obtaining insight into the beauty and harmony of God's universe. Similarly, the modern scientist uses the Mariner spacecrafts to determine conditions on Mars or Venus, but the idea that extrasensory perception (ESP) might be used to discover something about these planets is deemed absurd and ridiculous, if not insane. I do not propose a defense of ESP, but an existential approach to perception and learning theory gives a new perspective with which to view science. Specifically, for modern astronomers to assume that their "objective" science reveals the true nature of reality is just as pretentious and conceited as assuming that the earth is at the center of the universe.

We find that we are all easily seduced by modern science. After all, it works. Astronomers can calculate when the sun will rise tomorrow, and we find that at the prescribed instant the dazzling solar disk appears above the unobscured eastern horizon. Pioneer 10 sends back photographs of Jupiter, and the Apollo astronauts land on target on the moon and return safely.

Modern science is clearly mechanistic, not in the crude sense of gears and levers, but rather in terms of what is considered relevant. The only portions of reality considered truly important in modern science are those which can be measured and recorded by machines. The use of mechanical instruments such as telescopes, spectroscopes, galvanometers, and photographic film has the obvious effect of including or excluding precisely those portions of the total possible world-experience prescribed by the methodology of modern science. The inherent assumption that the true nature of reality is only that which can be recorded by machines reduces the modern scientist to the role of a "one-eyed, colorblind onlooker."

The economies of all modern industrial states are rooted in a technology founded on science. In view of the approach that modern science offers, it is not surprising that we now seem to be facing a variety of crises relating primarily to the quality of the environment and the ecological considerations imposed by our realization that the earth is a closed system. We live with the strange paradox of a technology that allegedly produces wealth and prosperity while actually devastating and impoverishing great portions of the earth's surface for generations to come.

If we are to continue on this planet as a viable species, we must

4

formulate new paradigms on which a postindustrial science can be developed. These new directions in science must have the precision and predictive power to which we have become accustomed, yet must be permeated with an awareness of the interrelationships between ourselves and the universe, between the microcosm and the macrocosm. In a postindustrial society we can no longer afford the luxury of ignorance concerning the directions of science. The average citizen cannot afford to sit back, cajoled by the professional elitism of the scientist, and be uninterested in the future course of science. Conversely, the professional scientist must recognize his or her moral responsibility to inform the public, lest in the face of aggressive technology we find ourselves impotently trying to cope with the dubious "benefits" and ill-conceived applications of pure research.

Perhaps in astronomy more than in any other physical science we find the best opportunity to develop this awareness. Over the centuries many of the most important and fundamental discoveries of physical science have come from studying the universe. For example, in the motions of the planets the scientist sees the laws of mechanics revealed in their purest and simplest form, unhampered by the friction and wind resistance encountered in the laboratory. It is no wonder, then, that Sir Isaac Newton was able to set the foundation of classical mechanics from his understanding of the workings of the solar system. Therefore, by examining the frontiers of modern astronomy we might well obtain insight into the future course of science.

In spite of the incredibly long history of astronomical observations, in spite of the fact that astronomy is correctly termed the "oldest science," it is impressive to notice how recent many ideas about the nature of the universe really are. Indeed, most of the concepts contained in any book on modern astronomy are less than a century old, and many of the topics currently discussed by professional astronomers were totally unknown one or two decades ago. For example, up until the middle of the last century astronomers had no firm idea as to precisely where the stars in the sky really are. Of course the apparent places of the stars have been known for thousands of years; star charts and maps have been produced by almost every civilization and there is evidence that the Cro-Magnons attempted to draw constellations on the ceilings of caves in southern Europe. But direct determination of the actual distances to the stars was delayed until the mid-nineteenth century. Astronomers were well aware that if stars are really as bright as the sun, then they must be incredibly remote because they are so dim in the nighttime sky. But

5

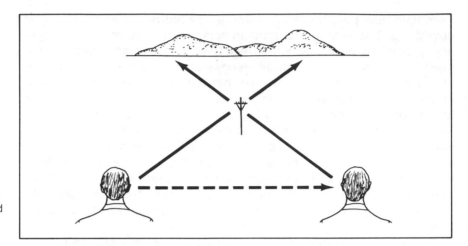

FIGURE 1-2. *Parallax and Common Sense.* The apparent location of a nearby object (telephone pole) with respect to a distant background depends on the location of the observer.

measuring these enormous stellar distances entailed severe observational difficulties.

A method by which distances to the nearest stars could be measured had been known for many centuries. This method, which involves the *parallax* of stars, is rooted in common experience. View a nearby object such as a telephone pole against a distant background. As you change your viewpoint, the telephone pole appears to move with respect to the distant background (Figure 1-2). This same parallactic effect can be directly applied to the stars. Suppose astronomers were to photograph the same field of stars on two occasions separated by a few months. Because the earth is moving around the sun, the astronomers would have photographed these stars from two very different locations. If the field of view happened to contain a nearby star, this star would have appeared to shift its position in relation to the background of remote stars (Figure 1-3). There is an inverse relationship between the amount of shift and the distance to the star: the farther the star, the smaller the shift. By measuring this shift astronomers can immediately calculate the distance to the star.

This method of parallax plays an important role in astronomy. It is one of the few methods that provide a *direct* determination of stellar distances. Parallax therefore constitutes our first step beyond the solar system, and all subsequent steps are in one way or another based on this first step. In addition, most other methods of distance determination are indirect and involve assumptions that can be called into question. The only thing that can be questioned about parallax is the accuracy of the observations.

The central difficulty with this method is that stars are so far away

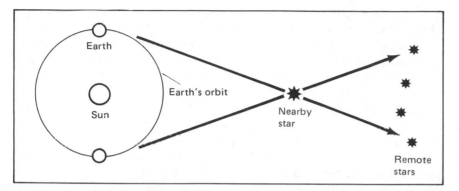

FIGURE 1-3. *Parallax and the Stars.* As the earth goes around the sun, a nearby star will appear to move back and forth with respect to the distant stars.

that accurate measurement of parallax is often difficult. It was not until 1838 that scientists finally had instruments capable of the necessary precision. At that time Friedrich Bessel of Germany measured the parallax of 61 Cygni. Shortly thereafter Thomas Henderson at the Cape of Good Hope and Friedrich Struve in Russia detected the parallaxes of α Centauri and Vega, respectively.

If someone were to ask you, "What is the distance, by air, from New York to Los Angeles?" you would have the option of a variety of correct answers. For example, you might respond by saying "155,295,000 inches." Of course this answer is very precise, but it sounds ridiculous if not humorous because the inch is not a convenient unit of measure in expressing the distances between cities on the earth. A more convenient unit of measure is the mile, and a more meaningful answer to the question of the distance between New York and Los Angeles is "2,451 miles."

When the distances to the stars finally became known, new units of measure had to be invented. To express stellar distances in miles would be just as ridiculous as expressing distances between cities on the earth in inches or millimeters. One of the most convenient yardsticks or units of measure invented to meet this need is the *light-year.* A light-year is simply the distance light, moving with a velocity of 186,281 miles per second, can travel in one year.* One light-year therefore equals about six trillion miles.

Using the light-year as our yardstick, we find that the distances to the stars can be expressed with very convenient numbers. For example, the nearest star, α Centauri, is about 4 light-years away. Sirius, the brightest star seen in the nighttime sky, has a distance of 9 light-years. Betelgeuse, the bright red star in the constellation of

---

* The speed of light is 2.99793×10^{10} cm/sec. For convenience throughout the text 186,000 miles per second will be used.

7

Orion, is 590 light-years away, while the distance to Rigel, the bright bluish star in the same constellation, is 880 light-years.

Although the light-year provides a convenient unit of measure in expressing stellar distances, we perhaps begin to wonder if nature is not trying to tell us something quite fundamental. For example, Aldebaran, the bright red star in Taurus, is 68 light-years away. This means that the light striking our eyes when we look at this star actually left Aldebaran 68 years ago. We are therefore *not* seeing this star the way it is now, but the way it was 68 years ago, before World War I. Similarly, when we look at more distant stars, we are actually seeing the way they appeared further in the past. When astronomers photograph a galaxy 90 million light-years away, the light that exposes their photographic film started journeying toward us when dinosaurs roamed the face of the earth (Figure 1-4). We therefore conclude that *as we look up into the nighttime sky we are actually*

FIGURE 1-4. *A Galaxy.* This photograph of a galaxy in the constellation of Eridanus was taken with the 200-inch telescope on Palomar Mountain. The light that exposed the astronomer's photographic film had been traveling toward earth for millions of years. (Hale Observatories)

8

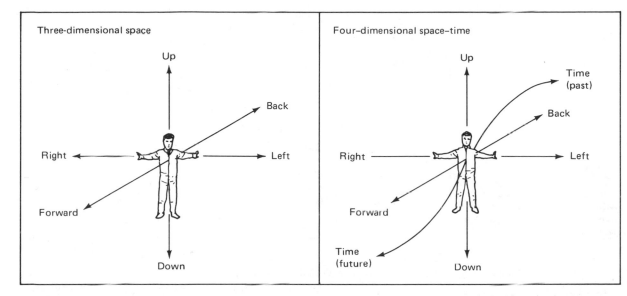

Three-dimensional space

Up

Back

Right ← → Left

Forward

Down

Four-dimensional space–time

Up

Time (past)

Back

Right → Left

Forward

Time (future)

Down

FIGURE 1-5. *Space and Space-Time.* As we are born, grow old, and die, we are three-dimensional creatures moving through a four-dimensional space-time.

looking backward into the past; as we gaze far out into space we are actually seeing backward in time. Indeed, we are drawn to the inexorable conclusion that time and space are intimately related, and that in order to understand the universe we must begin by appreciating the marriage of space and time into a continuum called *space-time.* We realize now that in viewing the star-filled sky, as so many of our ancestors before us have done, we are not only penetrating the three dimensions of space, but also the fourth dimension of time.

But what is really meant by space-time? As children we very easily learned to develop intuitive notions of the spatial and temporal dimensions. We learned that there are three spatial dimensions: forward-back, left-right, and up-down. As three-dimensional creatures in three-dimensional space, we have freedom of movement in any direction. But as the hours, months, and years pass by, we are also traveling through time. As three-dimensional creatures, we do not have the freedom to move back and forth in the fourth dimension of time. We are born, we grow old, and we die; we are powerless to turn the clock back or speed up its ticking (Figure 1-5).

Perhaps the best way to illustrate what is meant by space-time is with a simple example. Imagine climbing aboard an airplane in Los Angeles. Your destination is Seattle, and your flight stops for one hour in San Francisco. You could draw a graph, as shown in Figure 1-6, which describes your flight. On the horizontal axis of the graph measure how far you have traveled. On the vertical axis record the

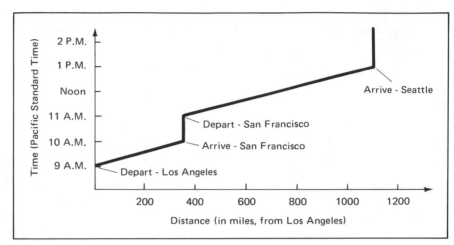

FIGURE 1-6. *An Example of Space-Time.* An airplane flight from Los Angeles to Seattle with a stop at San Francisco depicted in a two-dimensional space-time diagram.

time on your wristwatch. You depart from Los Angeles at 9:00 A.M. and arrive in San Francisco at 10:00 A.M. After an hour's delay, your plane continues to Seattle, arriving at 1:00 P.M. Knowing the distances between these cities, you find that completing the graph is an easy matter.

In a two-dimensional space-time diagram, the axis labeled "distance" tells where you are in the space dimension, while the axis labeled "time" tells where you are in the time dimension. Your path in this space-time is shown in Figure 1-6. When you are not moving, your path rises vertically in the diagram because your location in space is constant but time continues. When traveling, your path rises at an angle toward the upper right because you are getting farther and farther from Los Angeles as time passes.

In a similar fashion we could draw a diagram of a three-dimensional space-time. Suppose you were to enter a room through a door, walk first to a lamp, and then to a chair. Your path in ordinary space would look like that shown in the left of Figure 1-7. To see what your path would look like in space-time, a three-dimensional graph must be drawn. On one axis of the graph measure how far you moved in the north-south direction. On the second axis measure how far you moved in the east-west direction. And, finally, on the third axis record how much time has passed. In this space-time the paths or *world lines* of the door, lamp, and chair are straight lines parallel to the time axis. This must be the case, because the door, lamp, and chair do not change their location in space as time proceeds. You, on the other hand, follow a path that takes you from the world line of the door to the world line of the lamp and then to the

10

world line of the chair. As shown in the graph on the right of Figure 1-7, your path rises higher and higher because, as you move about the room, time continues.

Although these examples might seem rather trivial, they illustrate in a very general way what is meant by space-time. We could construct a drawing of a four-dimensional space-time, but this is unnecessary. As will be seen in numerous examples throughout this book, if we really understand what is going on in two dimensions, we can always generalize to three, four, or more dimensions. This is precisely what scientists do. Theoretical physicists doing research in the theory of relativity do not have some special mystical insight into what four dimensions look like. Rather, they begin by examining a problem in two dimensions, where their intuition is readily applicable. If they have done things properly they can always generalize or extend results and equations to four-dimensional space-time. We shall see that this is the case when the meaning of "warped" space-time is discussed.

As you might have expected, although the two previous examples of space-time illustrate all the basic concepts, physicists prefer to draw space-time diagrams in a slightly different fashion. Figure 1-8

FIGURE 1-7. *A Three-Dimensional Space-Time.* A person enters a room, walks to a lamp, and then to a chair. The person's path in ordinary space (left) and in space-time (right) are shown.

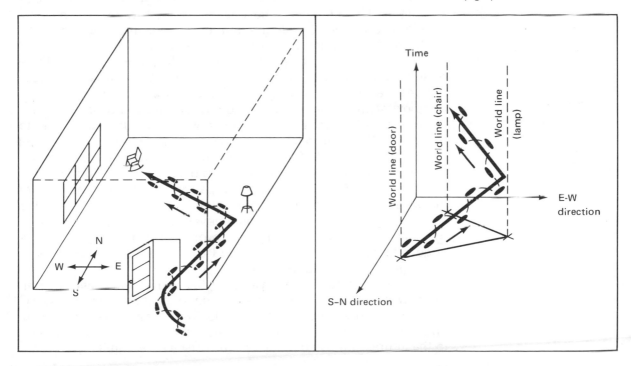

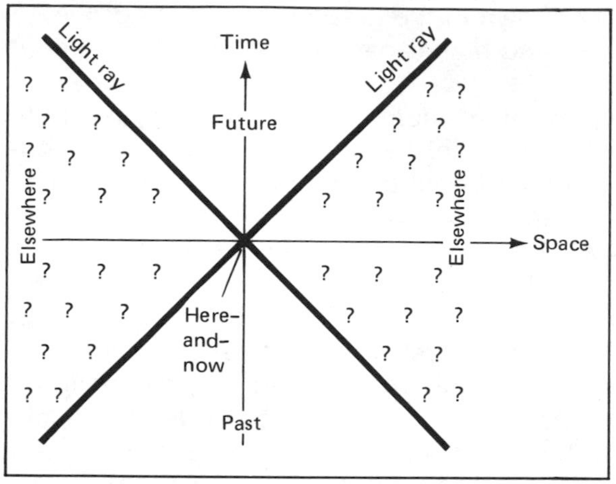

FIGURE 1-8. *Past, Future, and Elsewhere.* Physicists prefer to draw space-time diagrams in such a way that light rays travel along lines inclined at 45°. All of space-time then divides up into three distinct regions; the past, the future, and elsewhere.

illustrates this in a very general way. As in the case of the airplane flight from Los Angeles to Seattle, space is measured on the horizontal axis of the graph and time is measured on the vertical axis. These axes are, however, scaled in a very specific manner. If one inch vertically represents one second of time, then one inch horizontally is 186,000 miles, the distance light travels in a second. This results in light traveling along 45° lines in the diagram. If the center of the diagram is called *here-and-now* ("here" in space and "now" in time), all of space-time naturally divides up into three regions: *past, future,* and *elsewhere.* The boundaries of these regions are given by the paths of light rays passing through here-and-now at the center of the diagram. The meaning behind these three regions becomes apparent when we realize that, according to the special theory of relativity, it is impossible for anything to travel faster than the speed of light. For example, it would be easy to travel from here-and-now at the center of the diagram to the letter *t* in the word "future." A great deal of time passes and a small amount of distance is traversed; your speed along this trip is less than the speed of light. However, you could never get from here-and-now to the *s* in "elsewhere" because you would have to move across a large distance in a very short period of time and therefore have a speed greater than that of light.

Since it is impossible to travel at or faster than the speed of light, the only paths allowed in this space-time are those that are everywhere inclined at angles *less* than 45° with respect to the vertical. This means that starting from here-and-now you are always confined to the future. You can *never* travel into elsewhere. Similarly, in order

12

to be at here-and-now you must have come from somewhere in the past. For all time, those parts of the universe in elsewhere are forever forbidden to you.

From considering where you can and cannot travel in space-time, it is possible to define three basic kinds of trips, as shown in Figure 1-9. It is possible to travel from point *A* to point *B*. This path is inclined at an angle less than 45° with respect to the vertical; a great deal of time passes and not much distance is covered. Because more time than space is traversed, such a trip is termed *timelike*. The trip from point *C* to point *D* is inclined at an angle of exactly 45° with respect to the vertical. In view of how the space-time diagram was set up, for each second that passes 186,000 miles must be covered. The speed of an astronaut going from *C* to *D* would have to equal the speed of light and the trip is called *lightlike*. Finally, the path from point *E* to point *F* is inclined by more than 45° with respect to the vertical. Along this path, a great deal of space would be covered in a very short period of time. For this reason, the speed would have to be greater than the speed of light and the trip is called *spacelike*. Later in this book we shall make frequent use of space-time diagrams of the type discussed here.

All matter in the universe is restricted to timelike trajectories in the four-dimensional continuum of space-time. From the special theory of relativity and everything else known about the universe, it is totally impossible to accelerate matter to speeds up to or beyond the velocity of light. Therefore, lightlike and spacelike trips are always forbidden. To appreciate this more fully, central concepts in the special theory of relativity will be examined.

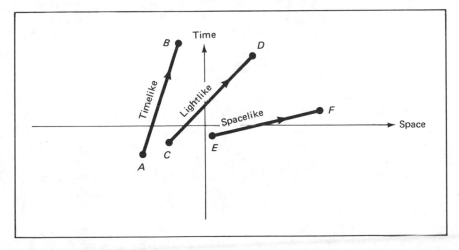

FIGURE 1-9. *Three Types of Trips.* Since it is impossible to travel faster than light, the motion of material bodies is restricted to timelike trips. Light travels along lightlike paths and spacelike trips are forbidden.

2 SPACE-TIME AND SPECIAL RELATIVITY

Think for a moment about history. Try to imagine the entire course of the past ten thousand years of humanity spread out before you, from earliest civilizations in the Indus and Euphrates valleys to Apollo astronauts driving around on the lunar surface. In this panorama a number of events, developments, and trends stand out as critically important or as having profound ramifications. Of course the events are the most obvious; the invention of the written word and the detonation of the atomic bomb are good examples. The trends and developments usually appear more gradually, as in the case of the decline of the Roman Empire or the remarkable (if not terrifying) population increases continuing through the twentieth century.

Looking around us today, we find that the discovery of the properties and uses of electricity ranks high on our list of history-shaping developments. In virtually every facet of our lives, in our homes and our offices, from communication to entertainment, electricity plays an all-important role. Of course, only a century ago this was not the case. Up until the 1800s, electricity had something to do with Leyden jars, Ben Franklin's kite, and shocks received in touching a doorknob after walking across a thickly carpeted room. It was not until the beginning of the nineteenth century that serious experimentation with electricity actually began in earnest. Perhaps the most important experiments revealing fundamental properties of electricity were those performed by Michael Faraday and Hans Christian Oersted. Almost by accident it was discovered that an electric current flowing through a wire produces a magnetic field! Indeed, if a pocket compass is placed near a wire in which an electric current is flowing, the compass needle is deflected from its usual northerly direction. Up until this time in history, electricity and magnetism were thought to be entirely separate phenomena. Electricity had something to do with lightning and magnetism dealt with the peculiar properties of certain iron rocks. Yet from the work of Faraday and Oersted it was apparent that these two phenomena are intimately related. Indeed, it became clear that a magnetic field is created anytime electric charges are set in motion, as shown in Figure 2-1.

FIGURE 2-1. *Electricity Produces Magnetism* (*Oersted's experiment*). When an electric current flows through a wire, a magnetic field appears around the wire, as shown by the direction of the compass needle before and after the switch is closed.

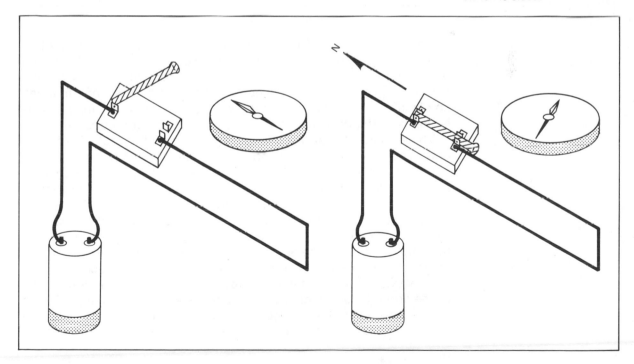

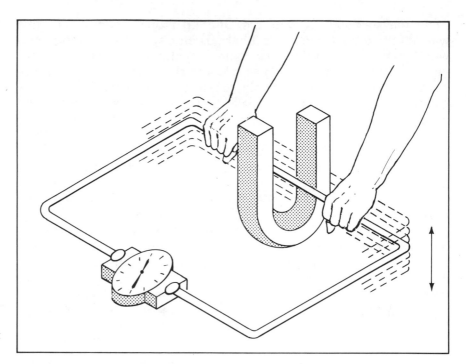

FIGURE 2-2. *Magnetism Produces Electricity* (*Faraday's experiment*). When a loop of wire moves through a magnetic field, an electric current flows through the wire.

Conversely, it was also shown in the early nineteenth century that changes, or motion, in a magnetic field set up an electric field even though no electric charges are present. For example, if a wire loop is moved back and forth between the poles of a horseshoe magnet, an electric current will flow in the wire, as shown in Figure 2-2. This is precisely the principle by which electric generators operate.

These fundamental discoveries inspired a great deal of research and experimentation, culminating with the work of the great Scottish physicist James Clerk Maxwell. In the eight years between 1865 and 1873, Maxwell succeeded in synthesizing all of the knowledge of electricity and magnetism into four simple equations. These four equations, which form the basis of *electromagnetic theory,* contain virtually everything known about the properties and interrelationships of the phenomena of electricity and magnetism. To accomplish the monumental achievement of unifying these phenomena, Maxwell had to take the revolutionary position that empty space around electric charges or magnets was dramatically affected by the presence of the charges or magnets. So-called "stresses," according to Maxwell, are created in empty space, giving rise to electric or magnetic *fields.* As a result, Maxwell's four equations (Figure 2-3) are often referred to as the *electromagnetic field equations.* This is the first time in the history of science that we find the concept of a field. Prior to the work of Maxwell, classical physics took the position that material bodies acted directly on each other at a distance rather than using empty space as an intermediary. Scientists now realize that the properties of space around objects can be altered by the presence of these objects.

In the mid-1800s, the remarkable discovery was made that Maxwell's four equations could be combined to give a *wave equation* describing the properties of light. This wave equation contained a number of surprises. First of all, it presented an entirely new way of thinking about light, namely, that a beam of light actually consists of simultaneous perpendicular oscillations of electric and magnetic fields, as shown in Figure 2-4. As a result, we can refer to light as *electromagnetic radiation.* The distance between successive peaks or valleys in the oscillations is called the *wavelength* of the radiation.

Second, it was realized that the electromagnetic wave equation did not place any restrictions or limitations on the wavelengths of the radiations it described. However, from experimentation physicists knew that ordinary visible light was restricted to a very small range of wavelengths. Therefore this wave equation actually predicted the existence of totally unknown types of electromagnetic radiation having wavelengths much longer and much shorter than ordinary light. In the decades following this theoretical prediction,

$$\nabla \times E = -\frac{\partial B}{\partial t}$$
$$\nabla \cdot D = \rho$$
$$\nabla \times H = J + \frac{\partial D}{\partial t}$$
$$\nabla \cdot B = 0$$

FIGURE 2-3. *Maxwell's Equations.* These four simple equations describe completely the all-encompassing relationship between electricity and magnetism.

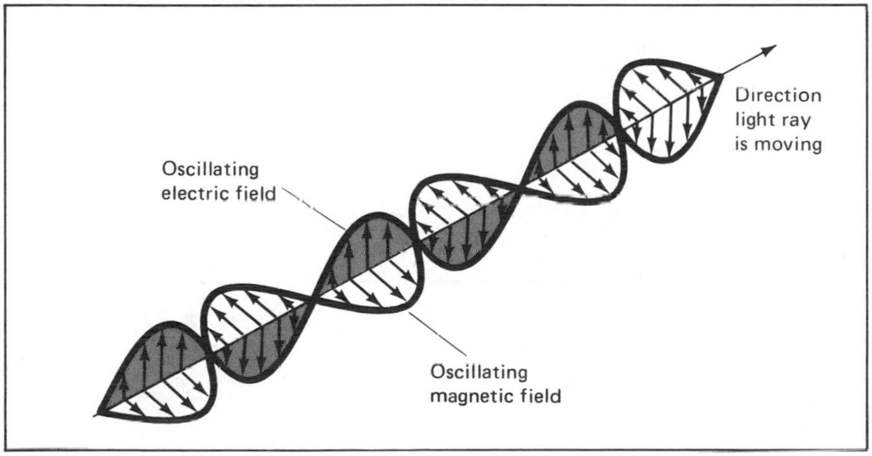

FIGURE 2-4. *Electromagnetic Radiation.* According to Maxwell's equations, all forms of light may be thought of as oscillating electric and magnetic fields.

Oscillating electric field

Direction light ray is moving

Oscillating magnetic field

many new types of electromagnetic radiation were discovered, and their names are now familiar. For example, radiations having wavelengths shorter than those of light include ultraviolet radiation, x-rays, and gamma rays, while at very long wavelengths we have infrared radiation and radio waves. All of these radiations, including ordinary light, comprise the *electromagnetic spectrum,* as shown in Figure 2-5.

Finally, one of the most perplexing properties of the electromagnetic wave equation is that, in its derivation from Maxwell's field equations, certain quantities combine to give rise to a number which, from experiments, has the value of 186,000 miles per second. In other words, embedded in the core of the wave equation is a velocity, usually symbolized by the letter *c*, which is recognized as being the speed of light. The implications of this unique development cannot be overemphasized. This is the first time in the history of science where, at the most basic and fundamental levels, a velocity appears in our description of the nature of reality. This appearance of the velocity *c* had ramifications for almost every concept about the universe, including intuitive ideas of space, time, and matter.

At first glance, the appearance of *c* in the wave equation simply says that all electromagnetic radiation must travel at a speed of 186,000 miles per second. But after a few moments' consideration we realize that we must ask "How?" and "With respect to what?" Sound waves are carried by the air and waves on the ocean are carried by the water, but what supports electromagnetic waves? To

FIGURE 2-5. *The Electromagnetic Spectrum.* Electromagnetic radiation ranges all the way from very short wavelength gamma rays to very long wavelength radio waves. Notice that visible light occupies only a very small portion of the spectrum.

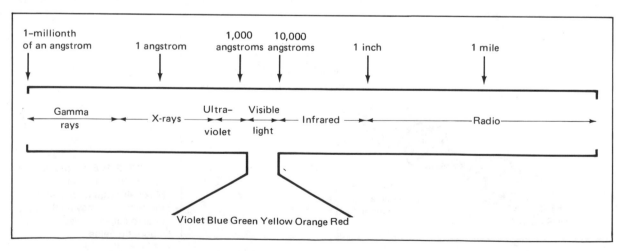

1–millionth of an angstrom   1 angstrom   1,000 angstroms   10,000 angstroms   1 inch   1 mile

Gamma rays — X-rays — Ultra-violet — Visible light — Infrared — Radio

Violet Blue Green Yellow Orange Red

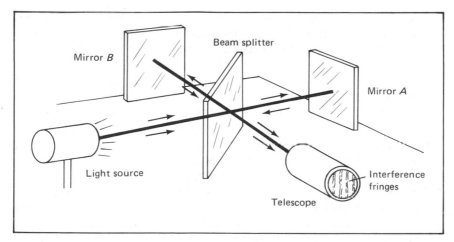

FIGURE 2-6. *The Michelson-Morley Experiment.* (Diagram of an interferometer.) Such an experimental setup was used by Michelson and Morley in an unsuccessful attempt to detect the earth's motion through the ether. The lack of positive results indicated that something was wrong with classical physics.

deal with this question, physicists of the nineteenth century postulated the existence of an all-pervasive medium called the "ether." This mysterious ether does not interact with anything in the material world; all the ether does is provide a carrier for electromagnetic waves. The logical inference, then, was that c is the speed of light in this mystical ether.

In the 1880s physicists realized that it should be possible to detect the earth's motion through the ether. After all, the ether must fill the universe, or how else could starlight ever get to us? Furthermore, the earth is going around the sun, so that at intervals separated by six months we should obviously be going in opposite directions in this universal bath of ether.

Two American physicists, Albert A. Michelson and Edward W. Morley, proposed a specific experiment by which our motion through the ether could be measured. A schematic diagram of their apparatus, called a *Michelson interferometer,* is shown in Figure 2-6. A source of light sends a beam toward the center of the apparatus. At the center of the apparatus is a beam splitter, which allows half the light to pass on through to mirror A while the other half of the beam of light is reflected at right angles toward mirror B. Great care is taken to ensure that the optical distances between the beam splitter and the two mirrors are equal, although at right angles. After being reflected by the mirrors A and B, the two beams return toward the center of the apparatus. Part of the beam from mirror B passes through the beam splitter and joins part of the beam from mirror A. The reunited beam is reflected toward a small telescope. It was well known, according to classical optics, that when these two beams come together on the last leg of their journey they will interfere with each other, producing *interference fringes.* These fringes are easily seen by looking through the small telescope.

The whole point of the *Michelson-Morley experiment* is that if the apparatus is left undisturbed the natural rotation of the earth over a 24-hour day will point the arms of the apparatus in various directions. If, for example, at 6 A.M. the path toward mirror *A* is parallel to the direction of the earth's motion and the path toward mirror *B* is perpendicular to the earth's orbit, then six hours later at 12 noon the two paths have reversed roles. In other words, at 6 A.M. the light going to and from mirror *A* is traveling parallel to the ether and the light going to and from mirror *B* is traveling perpendicular to the ether. By noontime, however, the light going to and from mirror *A* is now traveling perpendicular to the ether while the light going to and from mirror *B* is traveling parallel to the ether. This interchange in the roles of the two paths of light should produce very noticeable shifts in the interference fringes observed through the telescope. In this way, Michelson and Morley hoped to detect the earth's motion through the ether.

To understand this experiment more fully, imagine two good swimmers who swim with exactly the same speed in still water. Take your two friends to a river and have a race which begins from a dock, as shown in Figure 2-7. One swimmer will swim across the river and back (perpendicular to the river's current) while the second swimmer will swim an equal distance down the river and back (parallel to the river's current). If the river were still and there were no current, clearly the race would end in a tie. Simple arithmetic shows, however, that if the river is flowing, then the first swimmer (the one going perpendicular to the stream) will always win. It

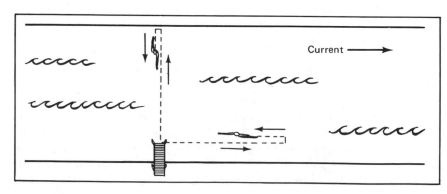

FIGURE 2-7. *Swimmers and a River.* Two swimmers who swim at the same speed in still water have a race. The one who chooses the path perpendicular to the river current will always win.

always takes a shorter time to swim back and forth across the river than it does to swim an equal distance downstream and return.

This same situation directly applies to the Michelson-Morley experiment. As soon as the light leaves the source in the apparatus, the light may be thought to have jumped into the river of ether flowing past the earth as the earth moves along in its orbit. From the previous analogy, the beam of light traveling perpendicular to the direction of the earth's motion always wins the race from the beam splitter to the mirror and back. However, since the earth is rotating, every six hours the winning and losing roles of the two paths are interchanged. It is this interchanging of the travel times of the beams of light which should produce the fringe shifts that Michelson and Morley expected.

This experiment was first performed in 1880, and to everyone's surprise no detectable shifting of the interference fringes whatsoever was observed. The conclusion is that either (1) the earth really is not moving or (2) the ether does not exist and there is something fundamentally wrong with the ideas about reality.

Although we have taken an experimental approach to the problems of the appearance of c in the wave equation, there are a host of theoretical difficulties. For example, consider a flashbulb such as used in photography. When a flashbulb is set off, a shell of light begins expanding outward in all directions. The person holding the flashbulb (the observer at rest with respect to the source) sees a *spherical* shell of light expanding from the source. However, an observer moving relative to the flashbulb should see an *ellipsoidal* shell of light radiating from the source. Something that is at the same time spherical and not spherical implies a paradox to the Western mind.

In 1905 a young German physicist working in a patent office in Switzerland succeeded in formulating a new and totally self-consistent theory about how we should look at reality. This theory, the special theory of relativity, was specifically designed to remove all of the difficulties associated with c in electromagnetic theory. Albert Einstein started off with the basic and far-reaching assumption that *the speed of light in a vacuum is an absolute constant.* In other words, anyone who measures the speed of light will *always* obtain the same answer, regardless of any relative motion between the source and the observer. Put another way, the speed of light is completely independent of the velocity of either the source or the observer.

This assumption is directly opposed to intuition and everyday experience. For example, imagine a person on top of a train moving with a speed of 50 miles per hour, as shown in Figure 2-8. This person throws a rock with a speed of 10 miles per hour in the direction in which the train is moving. According to an observer standing alongside the train tracks, the speed of the rock is 60 miles per hour (50 miles per hour for the train *plus* 10 miles per hour for the rock, relative to the train). This is common sense. Similarly, if this person on the train turns around and throws the rock with the same force toward the caboose, then the observer standing by the train tracks will see the rock moving at 40 miles per hour (50 miles per hour for

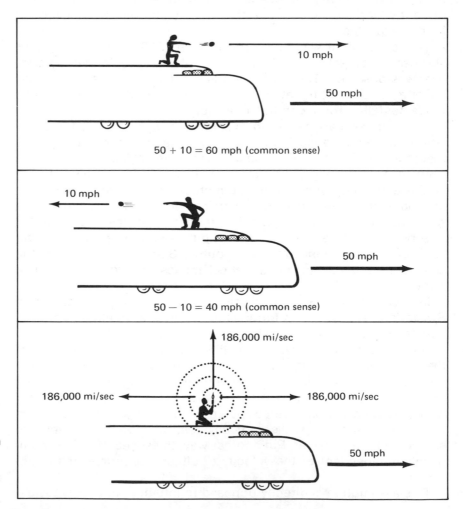

FIGURE 2-8. *Rocks, Train, and Candle.* Common sense dictates that the speed of a rock (relative to the ground) thrown by a person on a train depends on the speed of the train. However, the speed of light is independent of the velocity of the source.

22

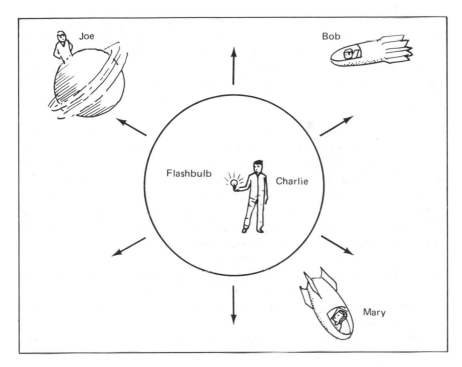

FIGURE 2-9. *An Expanding Shell of Light.* According to the assumption of the absolute constancy of the speed of light, all observers agree that they see an expanding spherical shell of light. But they will *not* agree on the rate at which their clocks measure time or on the lengths of their rulers.

the train *minus* 10 miles per hour for the rock relative to the train). This also is common sense. But if the person on the train turns on a light, both that person and the one standing on the ground must conclude that the light travels in all directions with the same speed of 186,000 miles per second, regardless of how fast or in what direction the train is moving. In order to come to this conclusion both people find that they must abandon many of their old, intuitive notions about space and time.

The framework of the special theory of relativity can be constructed directly from the assumption of the absolute constancy of the speed of light. Again imagine someone setting off a flashbulb. The person holding the flashbulb (Charlie in Figure 2-9) sees an expanding spherical shell of light moving away from him equally in all directions at 186,000 miles per second. In view of Einstein's assumption, *everyone* who watches this flash of light must agree that it is moving outward at 186,000 miles per second. In other words, everyone (Bob, Joe, and Mary in Figure 2-9) sees an expanding *spherical* shell of light. In order for all observers—whether moving or not—to see a spherical shell, conventional ideas about the nature of clocks

23

and rulers must be abandoned. Specifically, by requiring that any two observers moving relative to each other *both* observe a spherical shell of light, it is found that their rulers and clocks do not agree. Each says that the other person's clocks are slow and that the rulers measure different lengths in different directions.

At the heart of the special theory of relativity are a set of equations called the *Lorentz transformations.* These equations tell us how various things appear to different observers moving relative to each other. For example, one of the predictions of the theory is that, according to a stationary observer, moving clocks appear to slow down. This effect is sometimes referred to as *time dilation.* In other words, if you were in communication with an astronaut in a spaceship traveling through the solar system at a high speed, you would conclude that the clocks aboard the spaceship were all ticking too slowly. You, being at rest on the earth, would conclude that time has slowed down for the moving astronaut. This conclusion necessarily follows from the assumption that the speed of light is an absolute constant. If both you and the astronaut measure the speed of light and are to obtain exactly the same answer, according to you the astronaut's clocks must be slow.

Figure 2-10 graphically displays the Lorentz transformations for time. Specifically, this graph answers the question, "How long does one second on a moving clock appear to last according to a stationary clock?" For example, from this graph, if an astronaut is moving past you at 60 percent the speed of light, then 1 second measured by the astronaut's wristwatch will actually last $1\frac{1}{5}$ seconds

FIGURE 2-10. *The Dilation of Time.* The Lorentz transformations predict that time intervals measured by a moving clock will appear to last longer than the same intervals measured on a stationary clock.

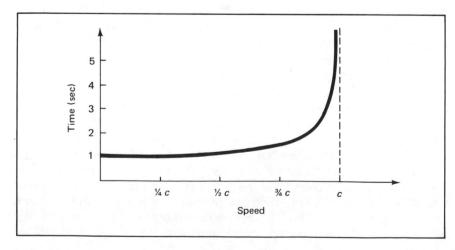

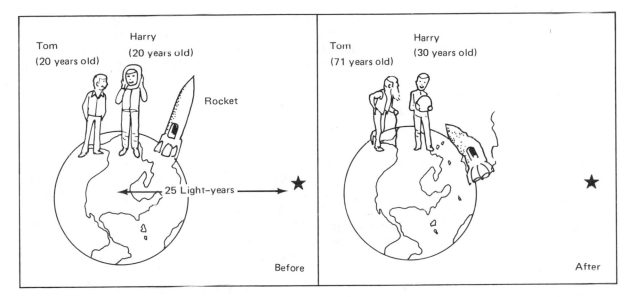

Before
After

FIGURE 2-11. *Tom and Harry.* Tom stays on the earth while Harry travels at 98% the speed of light to a star 25 light-years away and returns. Although they both start out being 20 years old, at the end of the trip Tom is 71 while Harry is only 30 years old.

according to your watch. In addition, from this graph, it is seen that the effect of the slowing down of time becomes strongly pronounced only at speeds very near the speed of light. Indeed, as the speed of light is neared, this dilation effect becomes infinitely large; at the speed of light, time seems to stop altogether.

The effect of the slowing down of time is responsible for a few conceptual difficulties and paradoxes. For example, in the ordinary world, if you go up to someone and tell her that her wristwatch is going slow, she could easily argue with you by saying that her watch was just fine but yours was going fast. Not so in relativity. If two astronauts are moving relative to each other at some high speed, each astronaut can assume that he is at rest and the other astronaut's clocks are going slow.

Along the same lines we might imagine what happens in the case of an interstellar spaceflight at very high speeds. For example, consider two young men on the earth, Tom and Harry, who are both 20 years old (see Figure 2-11). They have a spaceship capable of traveling at 98 percent the speed of light and are planning a round trip to a star 25 light-years away. Tom decides to stay home and Harry climbs aboard and proceeds on the journey. The astronaut travels at a constant velocity of 98 percent the speed of light, both out and back, and the total distance covered is 50 light-years. According to Tom, who stays behind on the earth, Harry's clocks have slowed

25

down. From the Lorentz transformations, one second on Harry's clocks lasts five seconds on Tom's clocks. Since Harry covers the entire trip of 50 light-years at a speed very nearly that of light, his journey according to the earth-based clocks takes 51 years. Therefore, at the conclusion of the star trek, *Tom is 71 years old.* On the other hand, since clocks have slowed during the trip, instead of 51 years having elapsed on the spacecraft, only 10 years passed by. As a result, upon returning to the earth *Harry is only 30 years old.*

This entire imaginary trip can be restated in such a way that each person takes the position that he is at rest and the other person's clocks go slow. Then it becomes unclear who ends up older than whom, resulting in the famous *twin paradox.* However, in the case of Tom and Harry it is obvious who is moving and who is not (after all, only one of them gets into the rocketship) and no confusion exists.

Just as in the case of time, the Lorentz transformations make specific statements about changes in distance and mass as a function of speed relative to various observers. In particular, according to a stationary observer, the rulers of a moving astronaut shrink when held in the direction of motion. This shrinking, sometimes referred to as the *Fitzgerald contraction,* becomes severe as you approach the speed of light (Figure 2-12). At the speed of light the rulers of an astronaut would theoretically shrink to zero length. Similarly, the masses of objects traveling at high speeds relative to a stationary observer appear to increase. At the speed of light a moving particle would theoretically appear to have infinite mass.

These predictions of the special theory of relativity have been

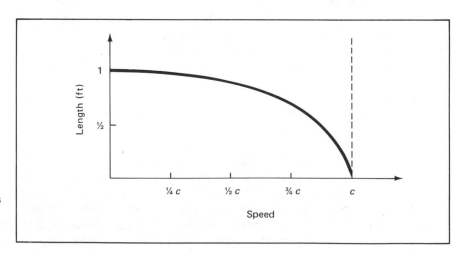

FIGURE 2-12. *The Fitzgerald Contraction.* According to a stationary observer, distances measured by a moving observer appear to shrink parallel to the direction of motion.

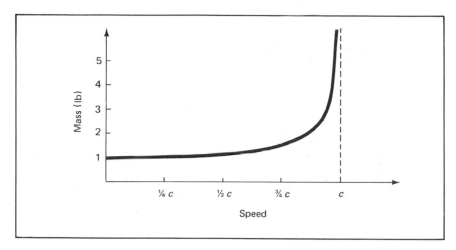

FIGURE 2-13. *The Relativity of Mass.* According to a stationary observer, the mass of a moving object appears to increase with increasing speed.

tested to a very high degree of accuracy in the laboratory by machines (cyclotrons, betatrons, synchrotrons, and so on) that accelerate particles to speeds very near the speed of light (Figure 2-13). It would be impossible to make any sense out of many experiments in nuclear physics if physicists did not take into account the effects of velocity on time, distance, and mass.

In addition, we now see physically why it is impossible to travel at speeds equal to or greater than the speed of light. Imagine, for example, a rocketship that never runs out of fuel. As the rocketship leaves the earth, it goes faster and faster. But as the spacecraft approaches the speed of light, the dilation of time begins to take over and, as seen by someone back on the earth, the rate at which the rocket's engines burn fuel starts to slow down. In fact, as the rocket gets very near the speed of light the engines appear to shut off. The effect of the slowing down of time is just sufficient to ensure that an astronaut never manages to burn those last few gallons of fuel to achieve those last few miles per second of speed to reach the critical velocity of light. Put another way, it would take you an infinite number of years working against the dilation of time to burn the necessary fuel to reach the speed of light. Any and all possible modes of propulsion will always face this insurmountable barrier.

While a single astronaut can never reach the speed of light, it might be thought that *two* astronauts could blast off from earth in such a way as to break the "light barrier." Imagine two astronauts traveling away from the earth in opposite directions, as shown in Figure 2-14. Suppose that each astronaut is moving away from the

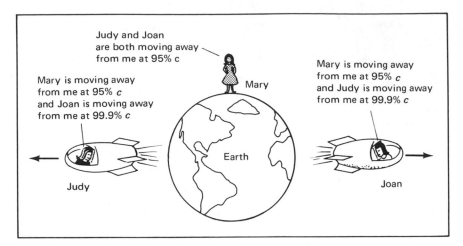

FIGURE 2-14. *A Scheme to Go Faster than Light.* Various schemes or tricks (based on common sense) to go faster than the speed of light all fail.

earth at 95 percent the speed of light. While everyone is agreed that the two astronauts are traveling away from the earth at 95 percent the speed of light, how fast are they moving away from *each other?* Common sense would suggest that the relative speed between the astronauts should be greater than 186,000 miles per second. However, when the techniques of special relativity are applied to this problem, it is found that common sense is again wrong. The Lorentz transformations for velocities dictate that the relative speed between the two astronauts is 99.9 percent *c.* The dilation of time is so effective that little "tricks" to get above the speed of light simply will not work.

In this connection it should be noted that many authors of popular science fiction have their heroes and heroines traveling in a special rocketship at speeds many times the speed of light. Similarly, advocates of flying saucers who believe that the earth has been visited by intelligent alien beings often discuss faster-than-light space travel. Such persons may not appreciate the implications of what they are saying.

The entire framework of modern physics is intimately related to consequences of special relativity. These consequences dictate that the speed of light is a barrier that cannot be exceeded under any circumstances. The counterargument put forth by science fiction authors and UFO buffs is that science may be wrong. Perhaps in decades or centuries to come, future scientists *will* discover new theories that will permit faster-than-light space travel. Although it is impossible to predict what science might be like a thousand years from now, we might try to appreciate some of the consequences of a

28

suprarelativistic theory. In particular, the speed-of-light barrier is such an integral part of modern science that any correct theory that allowed for supralight space travel would constitute a fundamental revolution in our understanding of reality. This revolution would be more profound and far-reaching than anything that has ever occurred in science. The gap in understanding between us and future astronauts capable of supralight space travel would necessarily be as great as the gap between a prehistoric man or woman and a modern nuclear physicist. It is totally presumptuous for people to assume that they could imagine what such astronauts might be like or how they might behave. The science fiction author whose characters travel faster than light can be compared to an ancient Egyptian poet trying to write a story about landing a Boeing 747 at Kennedy International Airport.

3 CONSEQUENCES OF SPECIAL RELATIVITY

The foundations of all physical science were dramatically revolutionized in 1905 with the publication of a small scientific paper, entitled "Zur Elektrodynamik bewegter Körper," by a young, virtually unknown physicist. In this historic paper, Albert Einstein succeeded in resolving any and all difficulties associated with Maxwell's electromagnetic theory. Specifically, Einstein reformulated physics in such a way that the basic laws of nature are the same for all observers, no matter how they might be moving relative to each other. The conviction that the laws of the universe must be the same for everyone is called the principle of *covariance.* When the mathematical equations of physics are written in such a way that they do not depend on the motion of the observer, these fundamental equations are said to be *covariant.* The price that must be paid for this elegant approach to physical reality is that certain basic quantities such as mass, time, and length are *not* the same for all observers.

To get a better idea of what is meant by a covariant formulation of electromagnetic theory, an example is in order. Imagine someone standing alongside an electrically charged metal ball, as shown in Figure 3-1. Charlie simply sees an electric field surrounding the metal ball, and he can measure the strength of this electric field with his scientific instruments. Now imagine a second observer traveling in a rocket past the first observer, as shown in the figure. According to Joe, the charged metal ball is moving relative to his rocket. An electric current, such as that flowing through the wires in your home, is simply electric charges in motion. Therefore Joe observes an electric current. But recall from the experiments of Oersted that electric currents produce magnetic fields. Thus the instruments on Joe's rocketship detect *both* electric and magnetic fields. Charlie sees only an electric field while Joe sees electric *and* magnetic fields. Furthermore, the strength of the electric fields measured by Joe and Charlie will be different. So it looks as though Joe and Charlie disagree as to what is really going on in this experiment.

Fortunately, however, Joe and Charlie have read Einstein's classic paper (or its translation, "On the Electrodynamics of Moving Bodies"). They learned that the strengths of electric and magnetic fields in the three dimensions of space (up-down, left-right, forward-back) can be combined into a single mathematical quantity called the *electromagnetic field tensor*. This new quantity is expressed in the

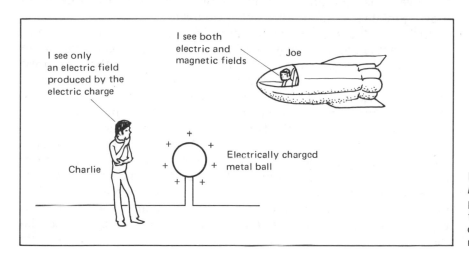

FIGURE 3-1. *Observers in Electrodynamics.* The same phenomenon involving electric and magnetic fields looks different to observers who are moving relative to each other.

$$\frac{\partial f_{\mu\upsilon}}{\partial x_{\upsilon}} = \mu_0 J_\mu$$

$$\frac{\partial f_{\upsilon\sigma}}{\partial x_\alpha} + \frac{\partial f_{\sigma\alpha}}{\partial x_\upsilon} + \frac{\partial f_{\alpha\upsilon}}{\partial x_\sigma} = 0$$

FIGURE 3-2. *Covariant Electrodynamics.* Electromagnetic theory can be formulated in space-time so that the equations are the same for everyone. Maxwell's four equations then reduce to only two equations that are said to be *covariant.*

four dimensions of space-time (up-down, left-right, forward-back, past-future). They also learned that electric currents and charges can be combined into a single four-dimensional quantity called the *four current.* When this is done, Maxwell's four equations (see Figure 2-3) reduce to only *two* equations which are completely covariant. These two equations are shown in Figure 3-2. They contain all the information that Maxwell's equations possess, except that all observers now unanimously agree that these equations correctly describe reality. There is no possibility for any disagreement between various observers, regardless of how they might be moving. Various bits and pieces of the electromagnetic field tensor correspond to the strengths of electric and magnetic fields in different directions. Various bits and pieces of the four current correspond to electric charges and ordinary current flowing in different directions. There will be disagreement between observers concerning the precise values of these bits and pieces, but there is no argument concerning the overall view when expressed using covariant formulation.

From the example with Charlie and Joe, it is seen that difficulties are removed and arguments are resolved when science is done in four dimensions. To illustrate further the power of Einstein's approach, consider space and time. As discussed in Chapter 2, different observers moving relative to each other will not be able to agree on distance or time measurements. Clocks slow and rulers shrink as the speed of light is approached. The distance between two objects is different for different observers. The time between two events is different for different observers. Can two observers moving relative to each other agree on *anything?*

Just as in the example of Joe and Charlie, it is possible to combine measurements of distances and time to give the *interval* in space-time between two *events.* Three of the pieces that comprise the interval come from measurements of the distances (up-down, left-right, forward-back) between the locations of the two events. The fourth piece comes from the amount of time that elapses between the occurrence of the two events. Different observers moving relative to each other will argue about specific distance or time measurements, but they all agree on the total interval in four-dimensional space-time. Such an interval is therefore said to be *invariant* because it is the same for everyone, as shown schematically in Figure 3-3. One observer might see two events occur close together in time (that is, nearly at the same instant) yet separated by a huge distance in space. A second observer might see the *same* two events

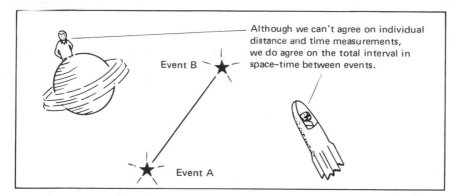

Although we can't agree on individual distance and time measurements, we do agree on the total interval in space–time between events.

Event B

Event A

FIGURE 3-3. *The Invariant Interval.* Observers moving relative to each other cannot agree on specific distance and time measurements between two events. Each observer can, however, combine his measurements of distance and time to give an *interval* between two events in space-time which is the same for everyone.

occur far apart in time (that is, separated by many hours) yet very close together in space. Nevertheless both observers will come up with the *same* total interval in space-time separating the two events. The shrinking of rulers and the stretching or dilation of time between the two observers as prescribed by the Lorentz transformations are of exactly the right amount that the interval is invariant.

While the interval between events in space-time is invariant, the Lorentz transformations relate specific distance and time measurements between observers. Perhaps the best way of displaying the effects of the Lorentz transformations is to examine what they do to space-time. The concept of space-time was introduced in the previous chapter, and from understanding the time-dilation effect we appreciate why lightlike and spacelike trips are prohibited to material particles. As usual, in drawings of space-time we shall adopt the convention of scaling our graphs so that light rays travel along 45° lines. For example, if one inch along the time axis represents one second, then one inch along the space axis equals 186,000 miles. To keep things straight, call the space-time of a stationary observer (such as us here on the earth) the *x, t system.* The space-time of a moving observer is to be called the *x', t' system.* The Lorentz transformations show what happens when we draw one space-time system on top of the other. As shown in Figure 3-4, the space-time of the *x, t* system looks like our familiar drawings of space-time. However, when the *x', t'* system is plotted on top of the *x, t* system (for convenience "here" and "now" coincide in both systems) the time and space axes of the *x', t'* system are tilted toward the 45° light-ray line. This tilting is symmetrical about the light-ray line only if the graphs are drawn so that light rays do indeed travel along 45° lines. Furthermore, the amount of tilting increases with increasing speed

33

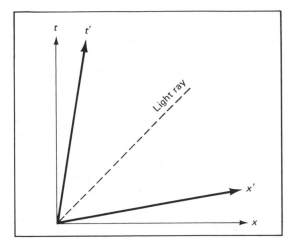

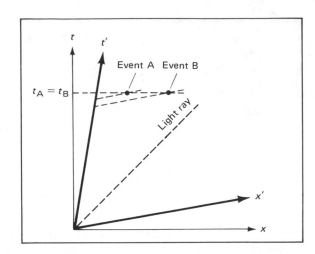

FIGURE 3-4. *The Lorentz Transformations* (*left*). The effect of these transformations on the space-time of a moving observer is to cause the space and time axes to be tilted toward the light-ray line.

FIGURE 3-5. *"Simultaneity" Is Nonsense* (*right*). Two events that appear to occur at the same time according to one observer can appear to occur at very different times according to another observer.

of the x', t' system relative to the x, t system. The greater the speed, the closer the x' and t' axes are to the 45° light-ray line. To obtain the location in space and time of an event in either of the two systems, we draw lines back to the axes, as shown in Figure 3-4.

From this way of displaying the Lorentz transformations it can be seen that terms such as "simultaneity" have no meaning. For example, consider two events, Event A and Event B, which appear to be simultaneous in the x, t system. By definition, this means that the two events occur at the same time and $t_A = t_B$, as shown in Figure 3-5. However, when the same two events are viewed from the x', t' system (which is in motion relative to the x, t system) these two events are not simultaneous. Indeed, the more distant event always appears to have occurred first.

Although it is impossible to go faster than the speed of light, and although the speed of light is an absolute constant, some unusual phenomena occur when we travel relative to sources of light. To appreciate some of these phenomena, imagine standing in a rainstorm holding an umbrella. Furthermore, suppose that there is no wind so that the raindrops are falling straight down. If you now begin to walk down the street, you obviously must hold the umbrella at an angle in front of you if you do not wish to get wet. The faster you walk, the further you must tilt the umbrella to prevent the raindrops from striking you, as shown in Figure 3-6.

A similar phenomenon occurs with starlight. The earth is traveling around the sun with an orbital speed of 18½ miles per second. Although this is a very small percentage of the velocity of light, the ef-

34

FIGURE 3-6. *Walking in the Rain.* A person walking in the rain must hold the umbrella toward the front in order not to get wet. The faster the person walks, the farther the umbrella must be tilted in order to stay dry.

fect of the earth's motion means that we must aim our telescopes slightly ahead of the locations of stars if the starlight is to go down the telescope tube. Just as you must tilt your umbrella in the direction you are walking, so also you must tilt your telescope through a small angle in the direction the earth is moving (Figure 3-7). This effect, called the *aberration of starlight,* was observed by James Bradley around 1725, when he noticed that there was a difference between the observed position and the true position of a star. This angle is always extremely small, never larger than 20½ seconds of arc. By comparison, the apparent angular size of Jupiter as seen from the earth is about 40 seconds of arc.

The aberration of starlight is so small simply because the earth is moving very slowly, at a speed only one ten-thousandth that of light. However, if you were to board a spaceship capable of traveling much closer to the speed of light, these aberration effects could become very noticeable.

Imagine a spaceship, like that in Figure 3-8, which has three large windows or portholes, one in the front of the spaceship, one in the rear, and one on the side. This spaceship is capable of traveling

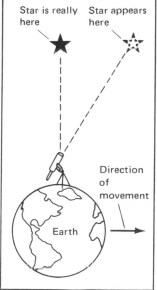

FIGURE 3-7. *The Aberration of Starlight.* Since the earth is moving, a telescope must be pointed slightly ahead of the star being viewed in order for the starlight to go straight down the telescope tube.

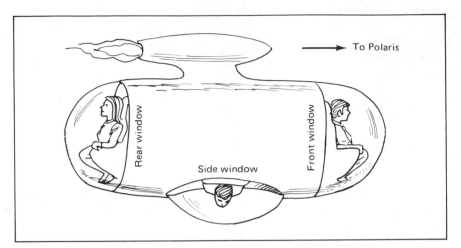

FIGURE 3-8. *A Relativistic Spaceship.* This futuristic spaceship can travel at speeds close to that of light. It has three windows, each of which permits a full 180° view of the sky. The astronauts can observe the stars from the front, the back, and the side of the spaceship.

at enormous speeds and through each window exactly one-half the entire sky can be seen. Furthermore, assume that you start on a journey toward the star Polaris, in the direction of the North Celestial Pole. Figure 3-9 shows what you would see out of the forward window. If the spaceship is not moving ($V = 0$), the constellations appear as they do on earth. At one-half the speed of light ($V = 50$ percent c), notice that the appearance of the sky is very distorted. The stars appear to crowd around Polaris and even some southern constellations have come into view. At 90 percent the speed of light, this distortion of the sky is so great that the constellation of the Southern Cross appears to be in front of the spaceship. Stellar images are being squeezed together at exactly the location toward which you are traveling. As a result, a bright starlike object begins to loom in front of the spaceship. Indeed, if you could achieve the speed of light, every star and galaxy in the entire sky would appear to be directly in front of the spaceship as you plunge toward an object of infinite brightness in an otherwise totally black sky.

Figure 3-10 shows what you would see out of the rear window of the spaceship. Again, $V = 0$ is the undistorted view. Closer and closer to the speed of light, however, fewer and fewer stars are seen. Stellar images disappear from view as they migrate toward Polaris. Of course, a star that has disappeared from view out of the rear window now can be seen out of the forward window.

This migration of stellar images toward the direction in which you are moving can be seen most easily from the side window, as shown in Figure 3-11. Note in particular how the Southern Cross appears to

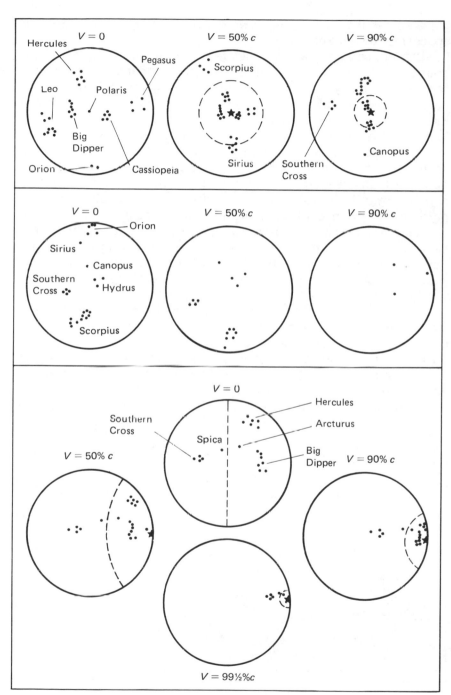

FIGURE 3-9. *View from the Front of a Spaceship.* The sky appears very different from a spaceship traveling toward Polaris (the North Star). As you get near the speed of light, all the stars in the sky appear to crowd together at the North Star. (Based on calculations by Drs. G. D. Scott and H. J. van Driel.)

FIGURE 3-10. *View from the Back of a Spaceship.* As an astronaut traveling toward Polaris gets close to the speed of light, he notices that very few stars are seen out his rear window. (Based on calculations by Drs. G. D. Scott and H. J. van Driel.)

FIGURE 3-11. *View from the Side Window of a Spaceship.* At higher and higher speeds, all the stars in the sky appear to converge on Polaris, toward which the spaceship is traveling. (Based on calculations by Drs. G. D. Scott and H. J. van Driel.)

37

move across the field of view toward Polaris as you accelerate your spaceship toward Polaris.

Although complex calculations are necessary to produce the star charts in Figures 3-9, 3-10, and 3-11, the fundamental ideas are rooted in everyday experience. Imagine driving down a freeway at 55 miles per hour. Furthermore, suppose it begins raining. Even though there is no wind and the raindrops are falling straight down, the rain pelts the windshield while the rear window on your car remains almost completely dry. As seen by you inside the moving car, it seems as if the raindrops are coming toward you. In the same sense, during a high-speed spaceflight it seems as though the starlight is coming toward the front window of the spacecraft. Stars from all parts of the sky appear to be in front of the spacecraft. Indeed, if future astronauts were to journey away from the sun at a velocity very near the speed of light, the sun would appear through the *front* window of their spacecraft. If they were to leave our galaxy, the galaxy would appear to be in front of them. The image of the sun or galaxy would, however, be very distorted.

In the same way that the positions of stars in the sky are affected by the motion of an observer, the colors of the stars are also changed. To see why this is so, again appeal to everyday experience. Imagine standing on a sidewalk as an ambulance speeds past with its siren wailing (Figure 3-12). As the ambulance approaches you, the pitch of the siren seems high. This is because the sound waves from the siren are crowded together in front of the ambulance. But when the ambulance is moving away from you, the siren's

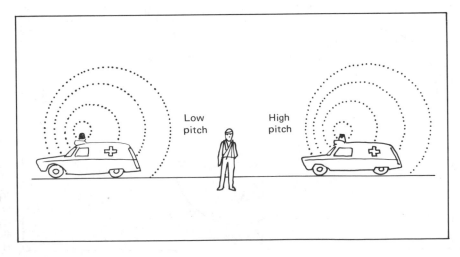

FIGURE 3-12. *The Doppler Effect.* The siren of an approaching ambulance seems to have a high pitch because all the sound waves are crowded together. Conversely, the pitch seems much lower as the siren recedes because the sound waves are spread out.

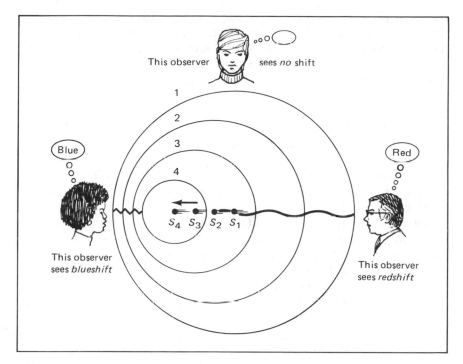

This observer sees *no* shift

Blue

This observer sees *blueshift*

s_4 s_3 s_2 s_1

Red

This observer sees *redshift*

FIGURE 3-13. *The Doppler Effect.* The light from an approaching source is blueshifted because the light waves are bunched together. The light from a receding source is redshifted because the light waves are spread out.

pitch seems much lower. Behind the receding ambulance the sound waves are stretched out, as shown in Figure 3-12. This phenomenon is called the *Doppler effect.*

The Doppler effect also occurs with light. As shown in Figure 3-13, the light waves from an approaching source of light are bunched together. The light therefore appears to have a higher frequency (or shorter wavelength) than usual. In the colors of the rainbow, blue light has the shortest wavelength. Therefore the light from an approaching source is said to be "blueshifted." On the other hand, if the source of light is receding, the light waves are spread out. The light appears to have a lower frequency (or longer wavelength) than usual. Of all the colors of the rainbow, red light has the longest wavelength and hence the light from a receding source is said to be "redshifted."

The exact amount by which the wavelength of a source is shifted directly depends on the relative speed between the source and the observer. If the speed is low, the shifting will be only slight. If the speed is very high, the shifting can be enormous. For example, imagine approaching an ordinary light bulb at 99.99 percent the speed of light. At such a high speed, the shift toward shorter wavelengths is so great that the light bulb seems to be emitting x-rays. Similarly if you were to move away from an ordinary light bulb at 99.99 percent the speed of light, you could detect only radio waves from the light bulb. In either case, although the light bulb really gives off only visible light, you could not see the light bulb with your eyes.

It is important to realize that although the wavelength or color of a light source depends on the relative velocity between the source and the observer, the speed of light is always the same. If you measure the speed of the radiation from either an approaching or receding source, you will always come out with the same answer: 186,000 miles per second.

Returning to the problem of observing the sky from a moving rocketship, we now realize that the Doppler effect has a profound influence on the colors of the stars. Those stars observed in front of the rocketship will have their light blueshifted since they are approaching the rocketship. Conversely, all of the stars seen out of the rear window of the rocketship will be redshifted since they are receding from the rocketship. This is true, however, *only* if the rocketship's speed is comparatively low. At high speeds—speeds that are a sizable fraction of the speed of light—a second effect is noticed.

Atoms that emit the light we observe can be thought of as little clocks. But recall that moving clocks appear to slow down. This slowing down of time is independent of the precise direction of motion. All clocks appear to slow down regardless of whether they are approaching or receding. Therefore, independent of the Doppler effect, the light from moving atoms will seem to have a lower frequency (or longer wavelength) simply due to the dilation of time. In other words, there are two simultaneous effects. In addition to the Doppler effect, the slowing down of time gives rise to a redshift. The dilation of time is less effective in shifting the wavelengths of light than the Doppler effect. Nevertheless, at speeds near the speed of light, *both* effects must be taken into account. Figure 3-14 shows those portions of the sky over which redshifts and blueshifts occur as observed from a rocketship traveling near the speed of light. As the rocketship approaches the speed of light, the redshift due to the dilation of time becomes more and more pronounced. As a result, the fraction of the sky over which blueshifts are observed gets smaller and smaller.

No discussions of special relativity would be complete without mentioning the strange subject of tachyons. While ordinary matter

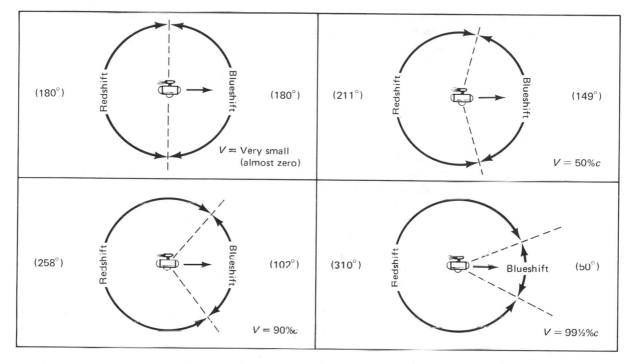

(180°) Redshift   Blueshift (180°) V = Very small (almost zero)

(211°) Redshift   Blueshift (149°) V = 50%c

(258°) Redshift   Blueshift (102°) V = 90%c

(310°) Redshift   Blueshift (50°) V = 99½%c

FIGURE 3-14. *Redshifts and Blueshifts.* The Doppler effect and the dilation of time result in substantial wavelength shifts in the starlight observed from a moving rocketship.

can never be accelerated to speeds equal to or greater than the velocity of light, the Lorentz transformations do allow mathematically for spacelike travel provided the "matter" making such trips possesses some strange properties. In the case of the ordinary world we can speak of *proper mass, proper length,* and *proper time* in the following sense. Imagine holding a brick in your hands. The proper mass of the brick is the mass which you, at rest with respect to the brick, actually measure with a scale. The proper dimensions of the brick are the lengths which you, at rest with respect to the brick, actually measure with a ruler. If the brick were radioactive, if it were made out of uranium, the proper half-life of the radioactive matter in the brick is what you, at rest with respect to the brick, actually measure with a clock. These proper quantities describe the properties of matter in the so-called *rest frame* of the matter. With respect to moving frames of reference, different quantities will be observed as dictated by the Lorentz transformations.

If a mathematician stares long enough at the Lorentz transformations, he or she realizes that faster-than-light travel is allowed provided that the proper quantities describing the matter traveling in this fashion are all "imaginary." The word "imaginary" has a very precise meaning to the mathematician (it involves the square roots of negative numbers), and in the ordinary world around us nothing is ever described by imaginary numbers. However, hypothetical matter traveling faster than light always must have imaginary proper mass, imaginary proper size, and imaginary proper time. Such matter is made out of *tachyons,* a word coming from the Greek root meaning "swift." Tachyons always travel faster than the speed of light, just as particles in the ordinary real world, called *tardyons,* always travel slower than the speed of light. In between the world of tardyons and the world of tachyons is the domain of *luxons,* particles which travel at exactly the speed of light, such as photons and neutrinos. Just as tardyons are usually at rest in the real world and it takes a lot of energy and effort to speed them up near to the velocity of light, tachyons usually have infinite speed, and it takes a great deal of energy and effort to slow them down to velocities near that of light.

Perhaps one of the most severe objections to the existence of tachyons involves violations of "causality." We are accustomed to very specific relationships between "cause" and "effect" in the world around us. Things happen because something makes them happen, and the cause always precedes the effect. However, this is not necessarily so for tachyons.

To illustrate these difficulties with tachyons, consider a simple experiment whereby a tachyon is created at one point, travels a certain distance, and then is destroyed at another point. For example, suppose you have a gun that shoots tachyon bullets. When you pull the trigger, a tachyon leaves the gun, travels across the room, and hits the wall. Call the instant when and where the tachyon leaves the gun Event A. Call the instant when and where the tachyon hits the wall Event B. Since the tachyon travels faster than light, the bullet's path in space-time must be a spacelike trip. In a space-time diagram describing this experiment, the path from Event A to Event B must be inclined at an angle greater than 45° from the vertical, as shown in Figure 3-15.

Now suppose someone is traveling past you at a very high speed. He is very interested in your tachyon gun and carefully observes what you are doing. What does he see? First of all, recall that the Lorentz transformations cause the space-time axes of a moving observer to be tilted. Figure 3-16 shows the tilted space-time of the moving observer plotted on top of your (stationary) space-time. Figure 3-16 also shows the spacelike path of the tachyon from Event A (when you pull the trigger) to Event B (when the tachyon bullet hits

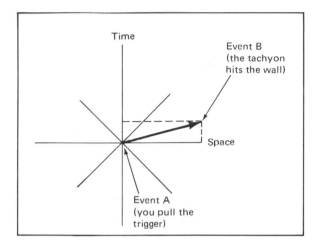

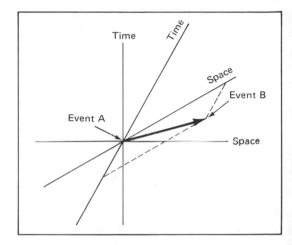

the wall). For convenience, you pull the trigger at the same instant that your moving friend passes by you. But, according to the moving observer, when does the tachyon hit the wall? To answer this question, simply draw a line from Event B parallel to the space axis of the moving observer back to the time axis of the moving observer, as shown in Figure 3-16. The resulting line intersects the time axis of the moving observer at a time *earlier* than Event A. According to the moving observer, Event B occurs *earlier* than Event A. The tachyon bullet hits the wall *before* it was shot from the gun!

On various occasions over the past decade scientists have done experiments to look for tachyons. Either tachyons do not interact very efficiently with matter of the ordinary world, or else they simply do not exist. These experiments usually involve looking for something quite unexpected in nuclear physics. For example, imagine an automobile accident in which a large truck traveling at a high speed slams into a parked Volkswagen. Obviously the small car will recoil from this collision by traveling some distance down the street in the *same* direction as the truck was originally moving. Only if tachyons are involved could the Volkswagen recoil in the opposite direction. In nuclear experiments light nuclei are bombarded by heavy nuclei. If a light nucleus were to be observed recoiling in a direction *opposite* to the original direction of the heavy nuclei, tachyons must be present. All laboratory experiments to detect the existence of tachyons have met with total failure.

Early in 1974, two physicists, Roger W. Clay and Philip C. Crouch, reported what might be called "suggestive evidence" for the exis-

FIGURE 3-15. *A Tachyon (left).* Tachyons, if they exist, travel along spacelike paths. This graph depicts a tachyon traveling from Event A (where it is shot from a gun) to Event B (where it hits a wall).

FIGURE 3-16. *Tachyons Violate Causality (right).* This graph shows the same spacelike trip of a tachyon from Event A to Event B as depicted in Figure 3-15. According to a moving observer, Event B occurs *before* Event A.

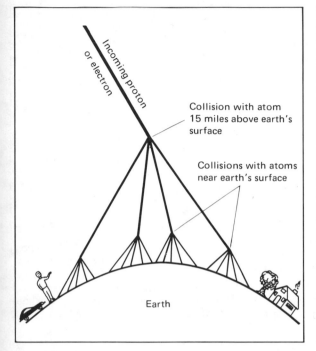

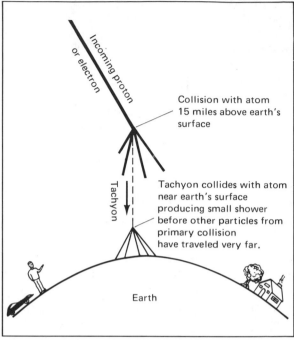

FIGURE 3-17. *A Cosmic Ray Shower* (*left*). Such a shower is produced by a high-speed proton or electron that shatters an atom high in the earth's atmosphere. About 60 one-millionths of a second later, all the pieces of the shattered atom "rain" down on the earth.

FIGURE 3-18. *Are Tachyons Produced by Cosmic Rays* (*right*)? If a tachyon is created by a cosmic ray shattering an atom high in the earth's atmosphere, it will reach the earth's surface ahead of all the other particles.

tence of tachyons. Their observations consisted of detailed analyses of "extensive air showers" produced by high-energy cosmic rays hitting the earth's upper atmosphere. Cosmic rays are high-energy nuclear particles (usually protons or electrons) coming from space at speeds nearly equal to the speed of light. As these particles, or primary cosmic rays, hit atoms in the earth's upper atmosphere, the resulting violent collisions produce a "shower" of nuclear particles. These showers are usually produced at altitudes of 15 miles and consist of many particles also traveling near the speed of light. These secondary particles collide with atoms at lower altitudes in the earth's atmosphere, resulting in a vast number of additional particles which "rain" down on the earth (see Figure 3-17).

It takes light about 60 one-millionths of a second to travel 15 miles. Therefore, ordinary (tardyon) particles in an extensive air shower *must* arrive later than 60 one-millionths of a second *after* the first collision of a primary cosmic ray with an atom high in the earth's atmosphere. In their experiments, Clay and Crouch discovered small showers immediately preceding the onset of an extensive air shower (see Figure 3-18). This would appear to indicate the exis-

44

tence of tachyons in the shower. They examined data from 1,307 air showers occurring from February through August of 1973. Their analyses indicate statistically significant numbers of particles arriving up to 60 one-millionths of a second before the beginning of main extensive air showers.

Although the work of Clay and Crouch does not constitute proof of the existence of tachyons, the results of their experiments are certainly intriguing. Hopefully, further observations of extensive air showers during the late 1970s will either confirm or disprove the existence of "precursors" occurring less than 60 one-millionths of a second before the onset of showers. If the existence of tachyons is confirmed, all of science will be dramatically affected. If tachyons exist, then causality can be violated; effects can occur before their causes. If tachyons exist, then the universe is irrational at a very fundamental level.

4 GRAVITATION AND GENERAL RELATIVITY

The primary problem facing ancient astronomers was to understand the motions of the planets. From observations spanning thousands of years, it was realized that the planets confine their motions to a band of twelve constellations around the sky known as the zodiac. Yet, from week to week and month to month, each planet would appear to trace out a complicated path among the fixed stars of these constellations (Figure 4-1). Indeed, the word "planet" comes from the ancient Greek verb that means "to wander."

There was great motivation in the ancient world to understand planetary motions. The planets were the embodiments of the gods, and to understand their motions would mean that people had fathomed the courses of the gods among the heavens. A priest or astrologer who could predict the motions of the planets among the fixed stars would obviously be in possession of great knowledge and power.

46

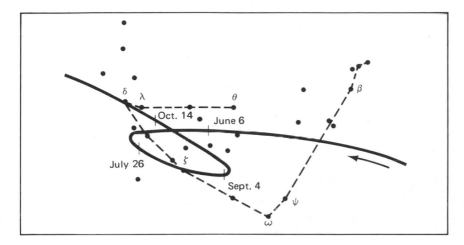

FIGURE 4-1. *The Path of Mars.* The apparent path of Mars through the constellation of Capricornus in 1971 is shown in this star chart.

Of all the cosmologies devised in ancient times to account for planetary motions, the most successful system was developed by Ptolemy, who lived in Alexandria during the second century A.D. Ptolemy's cosmology was a detailed elaboration on the earlier work of Hipparchus, which was based on two fundamental assumptions. First of all, ancient men and women believed that the earth is at the center of the universe. If you ride a horse or a chariot, or simply go for a walk, it is obvious that you are moving. But from the stillness of the ground below your feet, it seems reasonable to suppose that the earth is at rest. Thus, from observing the motions of the heavens from night to night, ancient people believed that the heavens revolved about our centrally located, immobile earth. Second, since the heavens were divinely created, they must be perfect. Since the heavens are perfect, any attempt to explain the motions of the planets must involve circles, which are the most perfect of all geometric forms. Therefore the Ptolemaic cosmology is a geocentric system in which everything must be explained by using circles and circular motion.

The prevalent theory in ancient Greece was that the wanderings of the planets among the constellations of the zodiac could be explained if one assumed that the planets revolve around *epicycles,* which in turn revolve around *deferents* centered approximately on the earth (Figure 4-2). Ptolemy's contribution was to work out all the details of this system. His *Almagest,* a collection of twelve books, contains everything the ancient astronomer or astrologer needed to know to compute the positions of the sun, moon, and planets on any night.

47

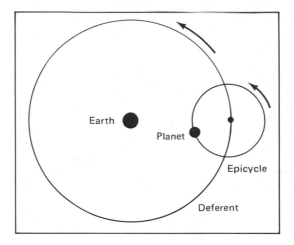

FIGURE 4-2. *The Ptolemaic System.* In Ptolemy's geocentric system, the motions of the planets are explained using circles. A planet is assumed to revolve about an epicycle, which in turn goes about a deferent centered on the earth.

For over one thousand years Ptolemy's cosmology survived as the true picture of physical reality. No other theory or set of ideas in astronomy has ever endured for this length of time. But, as the accuracy and precision of observations improved, it became painfully obvious that Ptolemy's system had to be modified to give the correct answers. Such modifications usually employed epicycles moving on the epicycles. Finally there were so many circles all rotating at various speeds that the entire geocentric cosmology was in danger of collapsing under the sheer weight of its own complexity. The time had come for a basic change in our thinking.

In the third century B.C., Aristarchus had proposed that everything would be a lot simpler if it were assumed that the sun was at the center of the universe and that the earth moved around the sun along with the other planets. That this novel system could account for the complicated paths of the planets is most easily seen from Figure 4-3. As the earth overtakes a slowly moving outer planet, this planet will appear to halt its usual eastward (direct) motion and move backward (retrograde) among the stars for a period of time, thus executing an apparent loop in the sky. Although this heliocentric cosmology would permit simpler explanations of planetary motions, it was rejected by ancient astronomers as unreasonable because it required the earth to be moving about the sun.

An important revival of this heliocentric cosmology occurred during the late Renaissance primarily due to the work of Nicolas Copernicus (Figure 4-4). Unlike Aristarchus, who merely proposed the general idea, Copernicus laboriously worked out all the mathemat-

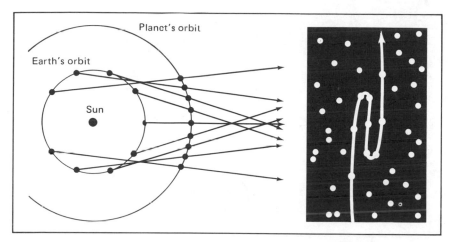

FIGURE 4-3. *The Heliocentric System.* The apparent motions of the planets can be explained easily in terms of the earth going around the sun. As shown here, when the earth overtakes a slowly moving outer planet, this planet appears to stop its usual motion and to go backward for a while.

ical details and proved that we can indeed use a heliocentric system to calculate the positions of the planets accurately. Unfortunately Copernicus continued to use circles in his work and, as a result, his description of the orbits of the planets contained epicycles. However, while the best Ptolemaic system required a total of 79 circles, Copernicus' heliocentric system contained only 34 circles to achieve the same accuracy.

In spite of opposition from the church, the heliocentric hypothesis gradually began to gain acceptance. Observational evidence came from Italy when Galileo Galilei, using the newly invented telescope, discovered the phases of Venus and the four largest satellites of Jupiter. Such observations showed that there were celestial objects *not* in orbit about the earth. For example, the only reasonable way of explaining the phases of Venus was to assume that it was orbiting the sun (see Figure 4-5).

Meanwhile in northern Europe, the young, brilliant astronomer Johannes Kepler (Figure 4-6) was experimenting with noncircular curves to account for the orbits of the planets. Kepler's teacher, Tycho Brahe, had compiled extremely accurate observations of the planets spanning two decades. In examining these records, Kepler concluded that any system utilizing epicycles had severe deficiencies. The stroke of genius that we should try different curves set the stage for modern astronomy.

After much trial and error, Kepler concluded that the motions of the planets could be very accurately described if we assumed that their orbits are *ellipses.* An ellipse is a curve that can be drawn very

FIGURE 4-4. *Nicolas Copernicus,* 1473–1543. (Yerkes Observatory)

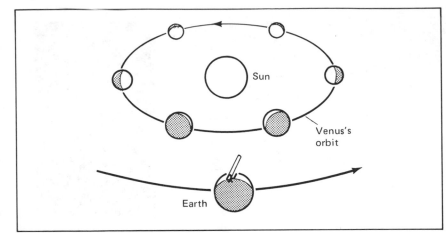

FIGURE 4-5. *The Phases of Venus.* Galileo discovered that Venus goes through phases like our moon. The appearance of Venus through a telescope can be understood in terms of Venus orbiting the sun, not the earth.

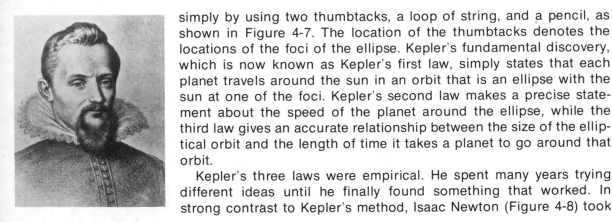

FIGURE 4-6. *Johannes Kepler,* 1571–1630. (Yerkes Observatory)

simply by using two thumbtacks, a loop of string, and a pencil, as shown in Figure 4-7. The location of the thumbtacks denotes the locations of the foci of the ellipse. Kepler's fundamental discovery, which is now known as Kepler's first law, simply states that each planet travels around the sun in an orbit that is an ellipse with the sun at one of the foci. Kepler's second law makes a precise statement about the speed of the planet around the ellipse, while the third law gives an accurate relationship between the size of the elliptical orbit and the length of time it takes a planet to go around that orbit.

Kepler's three laws were empirical. He spent many years trying different ideas until he finally found something that worked. In strong contrast to Kepler's method, Isaac Newton (Figure 4-8) took

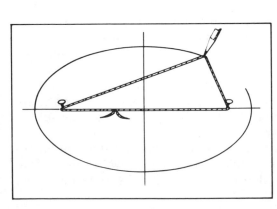

FIGURE 4-7. *The Ellipse.* An ellipse can be drawn with the aid of two thumbtacks and a loop of string.

an entirely theoretical approach to the problem of the motion of the planets. In the seventeenth century, Newton proposed three basic assumptions about the nature of physical reality. For example, his first assumption states that objects remain at rest or in motion in a straight line at a constant speed unless acted upon by an outside force. But the planets are not moving in straight lines. Therefore, there *must* be a force acting on the planets causing them to travel in elliptical orbits. By applying elaborate mathematical tools to Kepler's work, Newton was able to prove that this force is always directed toward the sun, and he was able to show how the strength of this force depends on the distance from the sun. The force is called *gravity,* and Newton's description of how the force behaves is formulated in his *universal law of gravitation.*

FIGURE 4-8. *Isaac Newton, 1643–1727.* (Yerkes Observatory)

Although Newton's law of gravitation is best expressed by a mathematical formula, the inverse-square dependence of the strength of gravity on distance can be illustrated by means of the graph of Figure 4-9. Suppose you are located one foot from an object that exerts a gravitational force on you of one pound. If your distance from the source of gravity is doubled, you would then experience a force only one-fourth as great. Similarly, at a distance of three feet, the force of gravity would be only one-ninth of a pound. If you moved closer to the source of gravity you would find that the gravity would increase. At a distance of half a foot, the force of gravity would be four times as great. At one-tenth of a foot, the force of gravity would be 100 pounds.

Another example of this classical behavior of gravity might be

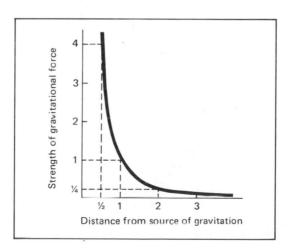

FIGURE 4-9. *Newton's Law of Gravitation.* This graph shows how the gravitational force of an object diminishes with distance from that object. Move twice as far away and the force is only one-fourth as great.

helpful. Imagine a person weighing 200 pounds standing on the surface of the earth, as shown in Figure 4-10. In round figures, he is 4,000 miles from the center of the source of gravitation. Now suppose this man climbed to the top of a ladder 4,000 miles high. He would then be twice as far from the center of the earth and therefore would weigh only one-fourth as much as before. If there were an ordinary bathroom scale at the top of this ladder, he would find that he weighs only 50 pounds.

It is important to note that this same result could be achieved by enlarging the earth to twice its original size. If the distances between all the atoms in the earth were doubled, the size of our planet would double. There are still the same number of atoms as before; we have not removed or added a single ounce of matter. All we have done is

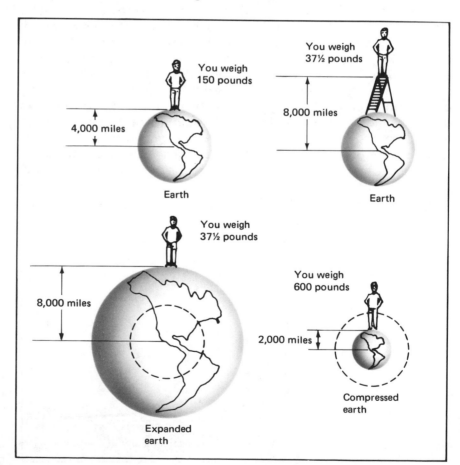

FIGURE 4-10. *The Behavior of Gravity.* How gravity works is illustrated in these four diagrams. The weight of a person depends on that person's distance from the earth.

redistributed the matter from which the earth is made. Then our friend, who originally weighed 200 pounds, would be located at a distance of 8,000 miles from the earth's center and he would now weigh only 50 pounds.

This behavior of gravity also works in the opposite direction. If the earth were squeezed down to half its original size, our friend would weigh four times as much, or 800 pounds. If the earth were compressed down to one-tenth its original size, the man standing on the surface would weigh ten tons.

It is clear that if objects could be compressed down to very small sizes, it would be possible to produce extremely intense gravitational fields. If a star, the earth, or even a grain of sand were in some way to collapse down to incredibly small dimensions, the strength of gravity at the surface of such an object might be so intense that not even light could escape. In 1795, the French mathematician Laplace noticed this interesting property of gravitation, namely that the escape velocity from a very collapsed or very massive object might exceed the speed of light. Yet it was not until 170 years later, until astronomers had understood many of the details of stellar evolution and seriously considered the implications of a violent and chaotic birth of the universe, that scientists began exploring the properties of the superintense gravitational field.

After having formulated his law of gravitation, Newton found that he could mathematically prove Kepler's laws as well as a great deal more. For example, using the mathematical tools he had developed, Newton showed that orbits of objects around the sun could be any one of a family of curves called *conic sections.* A conic section is any curve obtained by cutting a cone with a plane, as shown in Figure 4-11. Conic sections include circles, ellipses, parabolas, and

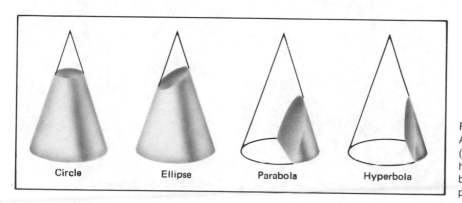

Circle   **Ellipse**   **Parabola**   **Hyperbola**

FIGURE 4-11. *Conic Sections.* A conic section is a curve (circle, ellipse, parabola, or hyperbola) which is obtained by cutting a cone with a plane.

hyperbolas. The exact orbit that an object will follow is determined by quantities such as the velocity of the object. With a relatively low speed an object might be confined to move along a closed orbit such as a circle or ellipse. But with a very high speed, the object may possess enough energy to escape from the solar system. In this case the object, such as a comet, would travel along a parabolic or hyperbolic path.

During the two centuries following Newton's pioneering work, numerous powerful and dramatic confirmations of his law of gravitation were made. For example, quite by accident in 1781 William Herschel discovered the planet Uranus in the constellation of Gemini (see Figure 4-12). After some observations, calculations of the orbit of Uranus were made in the best Newtonian tradition. However, by 1840 astronomers were aware that Uranus was *not* following its predicted path in the sky. Could it be that the law of gravitation does not work that far from the sun? Hardly. In England an astronomy student performed calculations that showed that Uranus' unusual behavior could be fully explained if there were a more distant planet exerting an unexpected gravitational force on Uranus. This additional force, although relatively small, would cause Uranus to deviate slightly from its anticipated orbit. Unfortunately, due to his lowly position, the student's calculations were ignored. Shortly thereafter a French astronomer independently performed the same

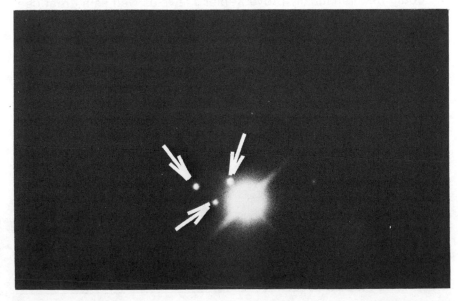

FIGURE 4-12. *Uranus and Three of Its Moons.* Uranus was discovered by accident in 1781. After several decades astronomers realized that the planet was not following its predicted orbit in the sky. (Lick Observatory)

54

FIGURE 4-13. *Neptune and Its Largest Moon.* Astronomers predicted the existence of Neptune in order to explain the unusual motions of Uranus. Neptune was really "discovered" with pencil and paper. (Lick Observatory)

calculations, which also predicted the location of a new undiscovered planet in the sky. A letter was sent to an observatory in Germany. The skies were clear on the day the letter was received and that night the eighth planet from the sun, Neptune, was first seen (see Figure 4-13). Newton's law of gravitation was so powerful and so universal that it could be used to predict the existence of undiscovered planets. Needless to say, a stormy debate between English and French astronomers ensued over who should receive credit for the discovery.

In spite of the successes of the law of gravitation, by the late nineteenth century it had become apparent that something was wrong with the orbit of the innermost planet, Mercury. Ideally, after all of the gravitational forces of the outer planets are subtracted, Mercury should be going around the sun in a perfect ellipse with the sun at one focus. Instead, the elliptical orbit of Mercury precesses very slowly for no apparent reason. Mercury actually traces out a rosette, as shown highly exaggerated in Figure 4-14.

Following the Uranus-Neptune example, some astronomers proposed the existence of a planet closer to the sun than Mercury, and everyone started looking for Vulcan. It does not exist. Other astronomers proposed modifying Newton's law slightly, but when they modified it to work for Mercury they could no longer explain the mo-

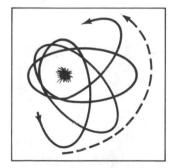

FIGURE 4-14. *Mercury's Orbit.* The orbit of Mercury very slowly moves around (or precesses), as shown in the figure. Newton's laws cannot explain this unusual abnormality.

tions of the outer planets. In short, nothing within the framework of classical Newtonian physics could account for this tiny but troublesome abnormality in the motions of Mercury. Once again the time had arrived for another basic change in our thinking.

In an earlier chapter it was explained that when we gaze at the stars in the nighttime sky we are really looking backward into the past. We are therefore compelled to think of time as a fourth dimension along with the usual three spatial dimensions. But on looking upward into the sky we also realize that on the astronomical scale, gravity is the most important force in nature. It is gravity that keeps the moon in orbit about the earth. It is gravity that holds the solar system together. It is gravity that dominates the interactions between stars and galaxies and perhaps dictates the entire past and future of the universe as a whole. Surely it would be exciting if, in some way, these two basic insights could be combined to produce a theory that might express one in terms of the other. Gravity would then be expressed in terms of the geometry of space-time, and the geometry of space-time in terms of gravity.

First of all it must be realized that we can actually dispose of the idea of gravity as a "force." Imagine standing in a small room with no windows. You notice that your feet are pressed firmly against the floor and the furniture and other objects in the room are also resting on the floor. Holding an apple out in front of you, you let go, and the apple falls directly toward the floor with a constant acceleration. Isaac Newton observing these phenomena would obviously conclude that the room is located on the surface of a planet, such as the earth, and all these effects are due to the force of gravity acting on the objects in the room. The force of gravity keeps you and the furniture on the floor; the force of gravity accelerates falling objects such as the apple. Although this may seem totally noncontroversial, shortly after the turn of the century Albert Einstein (Figure 4-15) proposed a completely different way of looking at this hypothetical room. Suppose that, unknown to you, this room were actually millions of miles out in space extremely far from any source of gravitation. Also suppose that, unknown to you, underneath the floor there were a powerful set of rockets with a very large supply of fuel. If the rockets had been turned on ever since you had been mysteriously placed in the room and if there were no noise or vibration from the rocket engines, they would be producing an acceleration of the entire windowless spaceship that would delude you into thinking that you were at rest in a gravitational field. This delusion would be so complete that no experiments whatsoever could possibly tell you

FIGURE 4-15. *Albert Einstein, 1879–1955.* (Yerkes Observatory)

56

Interstellar space

FIGURE 4-16. *The Principle of Equivalence.* From experiments performed in a windowless room, it is impossible to tell if you are at rest in a gravitational field or accelerating in empty space. The two situations are equivalent.

whether your room was at rest on a planet or attached to a spaceship (see Figure 4-16).

This example illustrates Einstein's principle of equivalence of gravity and acceleration. This *equivalence principle* simply states that in a small region of space ("locally") it is impossible to distinguish between acceleration and gravitation. This principle is used to dispose completely of the idea of gravity as a force.

There is a prevalent common misconception (of unknown origin) that the special theory of relativity does not apply to accelerated systems. Quite the contrary. Even at the fantastic accelerations experienced by high-energy nuclear particles, nuclear physicists accurately and routinely use the special theory of relativity to understand what is going on. Since the special theory of relativity is one of the best descriptions of physical reality scientists have, this theory can be used to understand the behavior of objects in our hypothetical windowless room. Indeed, we can barge ahead and use special relativity to solve all gravitational problems in this room because we can delude ourselves into treating gravity as a local illusion caused by acceleration. In fact, any gravitational field could be analyzed in this

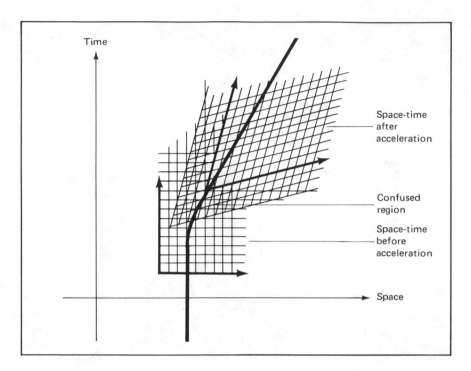

Time

Space-time after acceleration

Confused region

Space-time before acceleration

Space

FIGURE 4-17. *The Limitation of Flat Space-Time.* In examining accelerated objects, it is impossible to cover all of flat space-time with a single grid.

fashion. We could treat the gravitational field around a body such as the earth by breaking up all space into a large number of tiny rooms. In each room we think only of accelerations rather than gravity and apply the special theory of relativity. After having solved all the desired problems in each of these little rooms, we put all the pieces back together to obtain a comprehensible picture. In undertaking such a dissection and reconstitution we arrive at a generalization of special relativity. The final result is the *general theory of relativity.*

To see how this generalization is accomplished, consider the path of a person in space-time who undergoes a brief acceleration. Such a path is shown in Figure 4-17. Before and after the brief acceleration we can draw a space-time grid for this person. If he or she starts out at rest relative to the space-time diagram, the space-time grid of this person is identical with our own. However, after the acceleration, the person is moving relative to us with a particular velocity. According to the Lorentz transformations discussed in the previous chapter, the moving observer's space-time grid will appear to us to be squeezed slightly, as shown in Figure 4-17. There will be a region of confusion where the two space-time grids overlap, and we find that it is impossible to reconcile the two conflicting grids in this region.

Since gravity can be treated as being equivalent to acceleration in free space, the path of an object falling under the influence of gravity can be dissected into an infinite number of very tiny accelerations extremely close together. Before and after each one of these infinitesimal accelerations we can construct space-time grids, and we find that we are faced with an infinite number of regions of confusion in the overall space-time picture.

The source of this difficulty lies in the fact that special relativity is restricted to *flat* space-time. It is the rigorous application of the flat space-time at every point and at every instant that gives rise to the regions of confusion. If, however, space-time were allowed to be *curved,* the difficulties would disappear.

But what is meant by curved space-time? To answer this question, we must first have a very clear idea as to what is meant by anything being "flat" or "curved." As with any problem in relativity, it is extremely convenient to restrict the discussion to two dimensions. Then, provided the treatment is done properly, the final results and conclusions can be extended to four dimensions. In other words, if we understand precisely why the floor of a room is flat and a basketball is curved, we have the tools to understand what is meant by curved space-time.

Imagine a flat surface, as shown in Figure 4-18. Also imagine that a large number of ants start out from the same point on this surface. Each ant walks the same total distance r along the shortest possible path away from the common starting point. At the end of their trips, all the ants will be standing around in a circle that is centered on their starting point. In high school, geometry students are taught that the circumference of a circle is $2\pi r$. Therefore the total length of the curve joining all the ants at the end of their trips will be $2\pi r$.

Now imagine the same ants repeating this experiment on a surface that is *not* flat, as shown in Figure 4-19. Just as before, each ant walks the same total distance r away from the same starting point along the shortest possible path. In general, when the ants finish their trips, they will *not* be standing around in a perfect circle. Instead, they will be standing around along a curve that looks like a deformed circle. The total length of this resulting closed curve will *not* be equal to $2\pi r$.

The curvature of a surface is a measure of how much the total length of the "deformed" circle (i.e., the length of the closed curve connecting all the ants at the end of their trips) differs from $2\pi r$. If

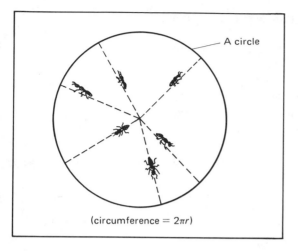
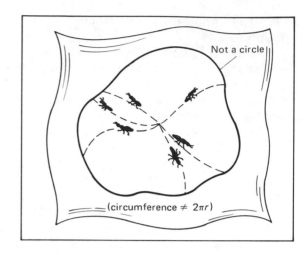

FIGURE 4-18. *Ants on a Flat Surface* (*left*). Each ant walks the same distance *r* away from the same point along the shortest possible path. The endpoints of their trips describe a circle whose circumference is $2\pi r$.

FIGURE 4-19. *Ants on a Curved Surface* (*right*). Each ant walks the same distance *r* away from the same point along the shortest possible path. The endpoints of their trips give a curve that is *not* a circle.

the length of the curve is *less* than $2\pi r$, then the surface is said to be *positively curved*. A basketball is an example of a positively curved surface. If the length of the curve is *greater* than $2\pi r$, then the surface is said to be *negatively curved*. A saddle is an example of a negatively curved surface. Only if the length of the curve is exactly equal to $2\pi r$ can the surface be called flat; it then has *zero curvature*.

The curvature of a surface may vary from one point to the next. In one region a surface might be flat; in another it might be positively or negatively curved. To cope with varying curvature, the mathematicians simply instruct their army of ants to walk only very short distances away from given starting points. The mathematicians thereby have the ability to measure the curvature of a surface at various locations.

This method of determining curvature can be extended to higher-dimensional spaces. To see how this is accomplished, realize that on a two-dimensional surface (that is, in "two-space") the ants moved away from a common starting point in all possible directions *on that surface*. At the end of their respective trips, the ants were standing along a curve that resembled a circle. In three-dimensional space (that is, in "three-space"), the ants again move away from a common starting point in all possible directions *in that space*. At the end of their respective trips, the ants will be standing on a closed surface that resembles the surface of a sphere. The curvature of the three-space is a measure of how much the surface of the resulting deformed sphere differs from $4\pi r^2$, which is the formula for the surface of a sphere in flat space. Similarly, in a four-dimensional space (that is, in "four-space") the ants leave a common starting point and move outward in all possible directions. At the end of their respective trips they will be standing on the surface of an object that could be called a *hypersphere.* The curvature of this four-dimensional space is obtained by comparing the three-dimensional "surface" of the hypersphere with the value this quantity would have *if* the four-space were flat.

In the nineteenth century, mathematicians such as Georg F. B. Riemann, Elvin Bruno Christoffel, and Gregorio Ricci developed the complete theory of curved spaces of any number of dimensions. The result of their efforts was a new branch of mathematics called *tensor analysis,* which involves mathematical quantities known as *tensors.* A mathematical quantity called the *Riemann curvature tensor* ($R_{\alpha\beta\gamma\delta}$) contains every conceivable piece of information about a curved space of any number of dimensions. From the Riemann curvature tensor, another mathematical quantity called the *Ricci tensor* ($R_{\alpha\beta}$) can be constructed which contains a large percentage of this same information. This is exactly what Einstein had been looking for!

The idea that gravity is a force can be discarded by thinking instead of local accelerations. The difficulties of applying special relativity everywhere to locally accelerated tiny rooms can be overcome by allowing space-time to be curved. Putting all this together, it would be incredibly beautiful if the gravitational field of an object could be considered as directly distorting the geometry of space and time. This is the central inspiration behind the general theory of relativity.

In a stroke of brilliant insight, Einstein realized that the gravitational field surrounding an object can be described as the curvature of space-time if the Ricci tensor is set equal to zero. The equation $R_{\alpha\beta} = 0$ tells us how much space-time is curved by the gravitational field of an object. This simple relationship is therefore called the *empty-space field equations.* The geometry of space-time around the earth or around the sun can be obtained by solving these equations. But inside the earth or inside the sun, space is *not* empty. To describe how space-time is curved where matter is present, Einstein developed a new set of field equations. Specifically, a quantity called the *Einstein tensor* ($G_{\alpha\beta}$) can be derived directly from the Ricci tensor. The full *Einstein field equations* are usually written so that on the left-hand side of the equal sign there are mathematical terms (the Einstein tensor $G_{\alpha\beta}$) which deal only with the geometry of space-time and so that on the right-hand side there are mathemati-

$$G_{\alpha\beta} = 8\pi T_{\alpha\beta}$$

FIGURE 4-20. *The Field Equations of General Relativity.* Gravitation is expressed in terms of the geometry of space-time by the Einstein field equations. Matter tells space-time how to curve and curved space-time tells matter how to behave.

cal terms (the *stress-energy tensor* $T_{\alpha\beta}$) which deal only with the physical properties of the matter that is the source of the gravitational field (Figure 4-20). In essence, by writing the Einstein field equations in this fashion, one makes geometry and matter equivalent. The central and most fundamental idea expressed by the field equations is that *the geometry of space-time tells matter how to behave,* and simultaneously *matter tells space-time how to curve.*

Consider a practical problem. Suppose you want to calculate how planets move about the sun. We solve the field equations for the empty space above the sun's surface and discover exactly how space-time is warped by the gravitational field of the sun. But where do we go from here? It is not enough to know all about the geometry of space and time due to the matter in the sun. We still do not know in which direction a planet might move.

To cope with this situation Einstein made the following simple assumption: an object moves along the shortest path in warped space-time. Such a path is called a *geodesic.* A geodesic is the generalization of the concept of a straight line in flat space. The equations that describe geodesics are called the *geodesic equations.* This formulation of a geodesic turns out to be extremely powerful. Freely falling particles and light rays all travel along geodesics. Therefore to solve the problem of a planet going around the sun, or any similar conceivable problem, all we do is

1. Solve the field equations. The solution tells us how space-time is warped.
2. Knowing the geometry of space-time, solve the geodesic equations. The solution tells us how particles or light rays move in this curved space-time.

At first glance nothing seems more beautiful and at the same time more ridiculous than particles traveling along geodesics. For example, imagine two people playing a game of tennis. Suppose the person hitting the ball toward his opponent lobs the ball high in the air. The tennis ball follows an arched path 25 feet above the court, finally arriving at the location of the other tennis player on the opposite side of the net, as shown in Figure 4-21. But, instead of lobbing the ball, the first tennis player could have hit a "line drive" toward his opponent 30 feet away. In this case, the tennis ball skims, just a few inches, over the top of the net, thereby making the total trip between players in a very short period of time. This second case is also shown in Figure 4-21. But what is going on here? In both

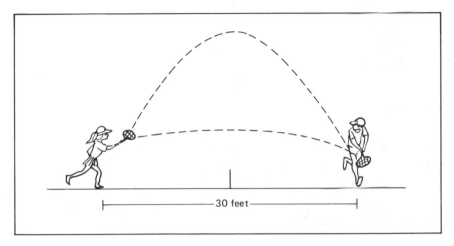

FIGURE 4-21. *A Tennis Game (in Space)*. The paths of tennis balls viewed in ordinary space look very different.

cases, the tennis ball starts from the same location and ends up in the same location. In both cases the tennis ball is falling freely along the entire length of its trip. But just take a look at Figure 4-21. The two paths are *very* different. So how could Einstein possibly claim that *both* paths are geodesics?

In the nineteenth century, Riemann became intrigued with the idea of expressing gravity in terms of the curvature of space. Yet, in spite of all his efforts, this gifted mathematician was totally unsuccessful because he spoke only of the curvature of *space.* Einstein, however, possessed the physical insight to formulate a geometrical theory of gravitation in terms of the curvature of *space-time.* In other words, the confusion associated with the previously described tennis game arises because the paths of the tennis balls were examined in space, not in space-time. To see what the tennis game looks like in space-time, a three-dimensional graph must be drawn. Along one axis we measure how far the balls move in the horizontal direction. Both balls cover a total distance of 30 feet on the ground. Along a second axis we measure the altitude of each ball along its path. The lobbed ball rises to an altitude of 25 feet while the line drive passes only a few inches over the net. Along the third axis we measure the time that elapses during the duration of the flights of the tennis balls. The lobbed ball takes a long time to travel between the two players while the line drive completes the trip in a much shorter period of time. The resulting graph is shown in Figure 4-22.

Careful examination of the paths of the tennis balls in space-time reveals that both paths are virtually identical. Both paths are very

63

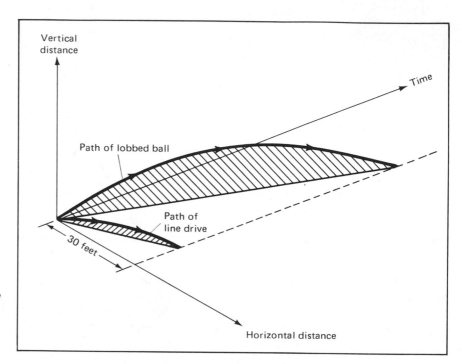

Vertical
distance

Time

Path of lobbed ball

Path of
line drive

30 feet

Horizontal distance

FIGURE 4-22. *A Tennis Game (in Space-Time)*. When the paths of tennis balls are viewed in space-time, their paths look identical.

nearly arcs of circles each having a diameter of about two light-years. While the paths of the tennis balls look very different in space, the paths in space-time have the same appearance. Of course, the line drive arrives at its destination sooner than the ball that was lobbed high in the air. Consequently, the path in space-time of the line drive is shorter than the path of the lobbed ball. Yet, both paths are pieces of the same circular curve. Both paths are the same geodesic.

This tennis game also illustrates another important point. A thirty-foot-long segment of a circle whose diameter is two light-years is almost a perfectly straight line. In other words, the geodesics of objects moving in the gravitational field of the earth are virtually indistinguishable from ordinary straight lines in space-time. This, in turn, means that space-time around the earth is almost perfectly flat. From the viewpoint of general relativity, the gravitational field of the earth is therefore *very* weak. Consequently, it is very difficult to perform any experiments on the earth—or anywhere in the solar system—which might be capable of detecting the very slight curvature of space-time. Testing the validity of general relativity poses a monumental task for the physicist or astronomer.

64

5 EXPERIMENTS IN GENERAL RELATIVITY

For two centuries the work of Sir Isaac Newton endured as the cornerstone in the unshakable foundation of classical mechanics. Thinking of gravity as a force could explain virtually everything. It was the force of gravity that held you in your chair. The force of gravity kept the moon in orbit about the earth. The force of gravity held the solar system together and dominated the interactions between stars and galaxies.

The successes of Newtonian mechanics mounted steadily over the years. In 1705 Edmund Halley published calculations concerning 24 comets. He noted that the orbits of bright comets seen in 1531, 1607, and 1682 were so similar that they might in fact be the *same* comet in a highly elliptical orbit about the sun. Elaborating on the work of Halley, Alexis Clairant predicted the return of this comet in 1758. It was sighted on Christmas night of that year, and was named *Halley's comet* (Figure 5-1). Using Newton's laws, a

FIGURE 5-1. *Halley's Comet.* Using Newtonian mechanics, astronomers of the eighteenth century discovered that this comet is a permanent member of the solar system. Halley's comet orbits the sun every 76 years, and is due to return in 1986. (Lick Observatory)

pencil, and some paper, astronomers had discovered a new permanent member of the solar system.

In the early nineteenth century, astronomers began discovering minor planets, or *asteroids,* orbiting the sun between Mars and Jupiter. Ceres was discovered by the Sicilian astronomer Guiseppe Piazzi on January 1, 1801. In March 1802, Heinrich Olbers discovered the second asteroid, Pallas. This was followed by the discovery of Juno in 1804 and Vesta in 1807. In all cases, their orbits conformed precisely to the predictions of Newtonian theory.

In the 1840s, John Couch Adams in England and Urbain Jean Joseph Leverrier in France independently concluded that observed deviations in Uranus' orbit could be explained by the existence of an eighth planet in the solar system. As noted in the previous chapter, their calculations led directly to the discovery of Neptune. Newtonian mechanics again emerged triumphant.

In spite of the multitude of successes of Newtonian gravitation, there was one problem. Beginning in 1859, Leverrier noticed that Mercury (Figure 5-2) was not exactly following its predicted orbit. As noted in the previous chapter, all attempts to account for Mercury's anomalous behavior within the framework of Newtonian mechanics were doomed to failure.

It should be stated that this irregularity in Mercury's behavior is very tiny. According to classical theory (that is, according to Newton, Kepler, and so on) the orbit of a single planet about the sun should be a perfect ellipse with the sun at one focus. However, there are other planets in the solar system besides Mercury. These other planets exert small gravitational forces on Mercury which cause its orbit to deviate slightly from a perfect ellipse. These outer planets

FIGURE 5-2. *Mercury*. In the mid-nineteenth century astronomers discovered that Mercury was not exactly following the orbit predicted for it by Newtonian theory. Although almost unnoticeable, this tiny anomalous behavior of Mercury cannot be explained within the framework of classical physics. (NASA)

are therefore said to *perturb* Mercury's orbit. From the mathematics of Newton's theory of gravitation, astronomers can calculate the exact size of these perturbations. That's not the problem. For many years it was well known that Mercury's orbit should precess due to the perturbations caused by all the other planets. However, the observed rate of precession is notably *larger* than the amount predicted from Newtonian theory.

To appreciate the dilemma facing astronomers of a century ago, consider a specific point along Mercury's orbit. For example, consider the point at which Mercury is closest to the sun. This point is called the *perihelion* and, as seen from the earth, has a particular orientation in the sky. Since Mercury's orbit precesses very slowly, the nearly elliptical path followed by the planet gradually changes its orientation. As a result, the direction to Mercury's perihelion moves slowly. This effect is so small that, over an entire century, the orientation of Mercury's perihelion precesses through only 1°33′20″, as shown in Figure 5-3. Of this observed rate of precession, 1°32′37″ per century can be explained as due to Newtonian effects. That leaves an excess of 43 seconds of arc that *cannot* be accounted for by classical theory. Although this discrepancy is very small, by the turn of the twentieth century it was clear that classical mechanics could not account fully for the behavior of the planet nearest the sun.

In 1916, Einstein proposed a radically new theory of gravitation called the *general theory of relativity*. According to this new theory, the gravitational field of an object manifests itself by warping spacetime. The stronger the gravitational field, the greater the curvature

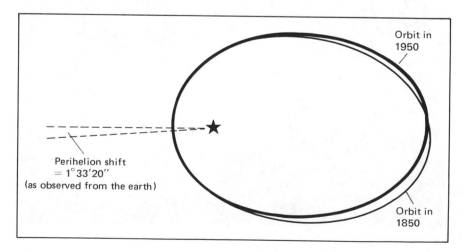

FIGURE 5-3. *The Precession of Mercury's Perihelion.* The position of Mercury's perihelion precesses through 1°33′20″ per century. Most (1°32′37″ per century) of this precession can be explained in terms of the perturbations by the outer planets.

Orbit in 1950

Perihelion shift = 1°33′20″ (as observed from the earth)

Orbit in 1850

FIGURE 5-4. *A View from Apollo 8.* Old-fashioned Newtonian theory is totally sufficient to calculate the orbits of astronauts to the moon and back. Any effects due to general relativity are too small to be noticed. (NASA)

of space-time. Particles and light rays travel along the shortest possible paths, geodesics, in that warped space-time.

In formulating a new theory of gravitation, Einstein realized that whatever ideas he came up with, his new theory *must* reduce to Newtonian gravitation in the *weak-field limit.* After all, Newton's ideas work pretty well. From the old-fashioned theory of gravity it is possible to calculate the orbits of comets and asteroids and to predict the existence of undiscovered planets. Even today, for calculating the trajectories of astronauts to the moon, only ordinary Newtonian theory is used (see Figure 5-4), because the gravitational field of the earth or the moon is *very* weak. In the language of general relativity, space-time around the earth or the moon is almost perfectly flat. We had some inkling of this in the previous chapter when we discussed a tennis match. We showed that the paths of tennis balls in space-time are really small pieces of huge circles. A 30-foot section of a circle whose diameter is two light-years is almost a straight line. Of course, Newtonian theory is totally adequate in explaining the trajectories of tennis balls in a tennis match. In other words, since Newtonian theory works so well in weak gravitational fields, Einstein realized that the field equations of general relativity must reduce to the equations describing Newton's law of gravity where space-time is almost perfectly flat. As scientists say, general relativity reduces to Newtonian gravitation in the weak-field limit.

When Einstein had succeeded in formulating the field equations of general relativity, he naturally wanted to apply his new theory to specific problems. The most obvious problem deals with the orbits of planets about the sun. According to Newtonian theory, the orbit of a single planet about the sun is an ellipse with the sun at one focus. But what does general relativity predict?

To begin, Einstein turned to the empty-space field equations. By solving these equations, he learned the shape of space-time surrounding the sun. Knowing the geometry of space-time, he then proceeded to solve the geodesic equations that told him how planets move in that warped space-time. The answer was *not* an ellipse! Instead, according to general relativity, the orbit of a single planet about the sun should be a slowly precessing ellipse. The orbit of a planet should precess quite naturally, without any perturbations from other planets. A precessing ellipse is simply the shortest path in the warped space-time about the sun.

When Einstein applied his discovery of precessing elliptical orbits to the planets in the solar system, he found that only in the case of Mercury should the effect be noticeable. Only in the case of the planet closest to the sun does the orbit lie in a region of sufficiently high space-time curvature that the relativistic precession could be detected. Specifically, in view of Mercury's distance from the sun, the rate of precession should be exactly 43 seconds of arc per century. This is precisely the amount of precession left unexplained by Newtonian theory. Finally, more than half a century after its discovery, the anomalous behavior of Mercury was understood. On December 15, 1915, Einstein wrote to a colleague in Poland:

I am sending you some of my papers. You will see that once more I have toppled my house of cards and built another; at least the middle structure is new. The explanation of the shift in Mercury's perihelion, which is empirically confirmed beyond a doubt, causes me great joy, but no less the fact that the general covariance of the law of gravitation has after all been carried to a successful conclusion.

The explanation of the precession of Mercury's perihelion was a major triumph for general relativity. For centuries, Newton and his law of gravitation had endured as the ultimate authority. Now there was a new theory of gravitation that worked even better. This new theory took the revolutionary approach of expressing gravity in

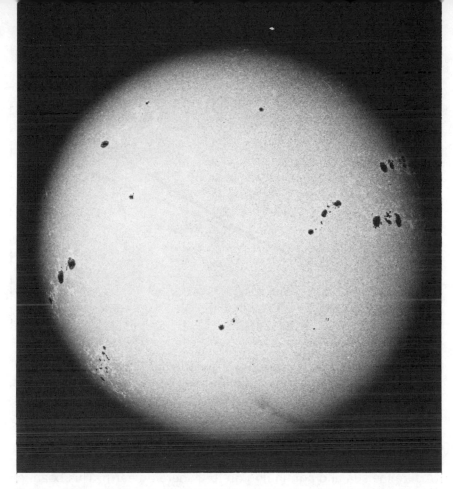

FIGURE 5-5. *The Sun.* The sun is the most massive object in the solar system. As a result, the strongest gravitational field in the solar system is located around the sun. (Hale Observatories)

terms of the geometry of space-time. No longer could gravity be thought of as a force. Instead, the gravitational field of a body warps space-time, and objects moving in that warped space-time travel along the shortest possible paths.

Whenever any new ideas or theories are proposed, scientists want to *test* these new concepts to see if they really are better than the old ideas. It is not enough for a new theory just to explain all the old, familiar observations and experiments. Ideally, the new theory should predict new phenomena that no one has ever thought of. Thus, around the time of World War I, Einstein began thinking of new observations that could be used to prove or disprove his general theory of relativity. It wasn't an easy job.

As noted earlier, everything involving gravity known to astronomers and physicists could be explained from the old Newtonian viewpoint, with the single exception of Mercury's precession. Everywhere in the solar system, space-time is almost perfectly flat, and thus thinking of gravity as a force is quite acceptable.

The sun is by far the most massive object in the solar system (Figure 5-5). Over 99 percent of the matter in the solar system is located

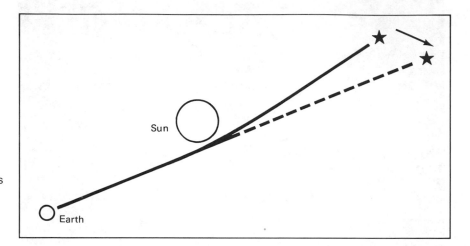

FIGURE 5-6. *The Deflection of Light by the Sun.* A beam of light passing near the sun's surface is deflected from a straight-line path by the curvature of space-time surrounding the sun.

in the sun. Since the sun is so massive compared to anything else, it has the strongest gravitational field of any object in the solar system and the most highly curved space-time in the solar system must be located around the sun's surface. Therefore the obvious place to look for the effects of general relativity is near the sun.

Every star in the sky sends at least a few beams of light near the sun's surface. For example, Figure 5-6 shows a beam of starlight grazing the sun and continuing on to us here on earth. This beam of light passes through the warped space-time around the sun. Since the beam moves along the shortest path in *curved* space-time, it is deflected from its usual straight-line trajectory. No one had ever seriously considered the possibility that gravity could bend light rays. According to Newtonian theory, gravity *cannot* affect the paths of light rays. Yet the idea of the gravitational deflection of light follows quite naturally from general relativity. After all, the path of a light ray must be curved if it passes through a region of curved space-time.

As in the case of Mercury's orbit, the effects of general relativity are very small. At best, a light ray just grazing the solar surface should be deflected through 1.75 seconds of arc. This is a very tiny angle. Beams of light that pass by the sun at larger distances should be deflected through even smaller angles since these beams are moving through a region of space-time where the curvature is less noticeable. Referring back to Figure 5-6, as viewed from earth, the observed position of a star near the sun should be shifted outward away from the sun through an angle of at most 1.75 seconds of arc.

72

You cannot see stars during the daytime; the sun is just too bright. However, during a total eclipse of the sun (Figure 5-7), the moon completely covers the blinding solar disk, and, for a few minutes, the stars come into view. By comparing photographs of the stars near the sun during a total eclipse with similar photographs taken months earlier (when the sun was in a different part of the sky) astronomers hoped to obtain a further test of the validity of Einstein's theory of general relativity.

Two expeditions of British astronomers were sent out by the Royal Society to observe the solar eclipse of May 29, 1919. One team traveled to Brazil and the other group was stationed on the west coast of Africa. To Sir Arthur Eddington, the leader of the African expedition, the first measurements of the photographic plates was the greatest event in his life. Einstein's prediction of the gravitational deflection of light had been confirmed beyond a doubt.

FIGURE 5-7. *A Total Solar Eclipse.* During totality, stars near the sun can be seen, although none appear in this photo, which was designed to reveal the solar corona. By accurately measuring the shifted positions of stars seen near the sun during a total eclipse, astronomers obtained an important confirmation of the general theory of relativity. (Hale Observatories)

73

During almost every eclipse since 1919, an astronomer somewhere has tried to measure the deflection of light by the sun. Because solar eclipses often occur over well-nigh inaccessible regions of the earth, astronomers who wish to observe a solar eclipse frequently find themselves lugging all their equipment up the Amazon River or across the Sahara Desert. And when the moment of totality finally arrives, the unfortunate astronomers may be standing knee-deep in a swamp disturbed by gnats and threatened by more dangerous creatures. In more scientific terms, the "experimental errors" in all such eclipse observations are often quite large. There must be a better way.

The need for more accurate tests of general relativity became quite critical during the late 1960s. At that time several very clever physicists had proposed new theories of gravitation which gained considerable notoriety. These new theories have many of the same features as general relativity—they all express gravity in terms of the curvature of space-time. However, the precise degree of warping of space-time in these theories differs slightly from the amount calculated according to Einstein's theory. For example, the most popular of all these non-Einsteinian theories was formulated by R. Dicke and C. Brans at Princeton University. In both Einsteinian *and* Newtonian gravitation, there is an important number called the *gravitational constant.* In a very general way, the size of this number tells how strong gravity is compared to other forces in nature. This number has been measured in laboratory experiments, and its value is listed as $G = 6.688 \times 10^{-8}$ dynes cm^2/g^2. In the late 1930s, however, the great physicist P. A. M. Dirac began to seriously wonder if the gravitational constant has always been exactly the same as it is today. He presented some intriguing arguments suggesting that perhaps in the distant past the gravitational constant had a very large value and has been gradually decreasing with time. Brans and Dicke followed up this idea and formulated a new relativistic theory of gravity in which the gravitational "constant" is *not* constant. The field equations of the Brans-Dicke theory look similar to those of Einstein's theory except that there are a lot of extra mathematical terms added on that allow for the changing value of the gravitational constant. The final result of the Brans-Dicke theory is that the deflection of light by the sun and the advance of Mercury's perihelion should be slightly less than the amounts predicted from Einstein's theory. Measurements of the deflection of light during total eclipses of the sun were not precise enough to distinguish conclusively between the two competing theories.

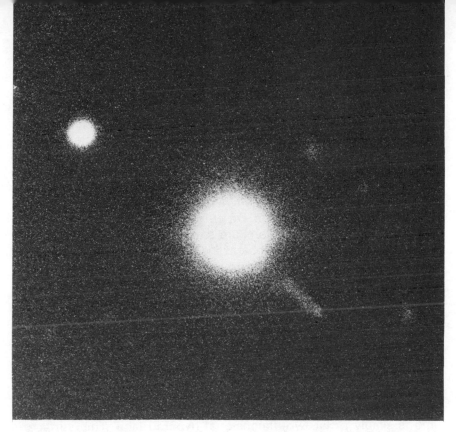

FIGURE 5-8. *The Quasar 3C 273.* Quasars are bright sources of radio waves. By measuring the deflection of radio light from 3C 273 by the sun, astronomers further confirmed the validity of the general theory of relativity. (Hale Observatories)

During the 1960s, astronomers discovered objects in the sky called *quasars*. At first glance, quasars look like ordinary stars, but upon closer inspection they have many of the properties usually associated with distant galaxies. Although the nature of quasars is not yet understood, quasars do emit large quantities of radio waves.

The fact that quasars shine so brightly in radio light suggested an important observation to radio astronomers. Every October 8 the sun passes in front of the quasar 3C 273 (Figure 5-8). As the sun approaches the location of 3C 273 in the constellation of Virgo, the radio light from this quasar should be deflected in exactly the same way that ordinary beams of light from the stars are deflected. Since the sun is comparatively dim in radio light, radio astronomers do not have to wait for an eclipse; the observations can be made in the comfort of a radio observatory.

During the early 1970s, radio astronomers carefully observed the deflection of radio waves by the sun. This was done by measuring the angular separation between the quasars 3C 273 and 3C 279 in October of 1972. As the sun approached the location of 3C 273, the angle between these two quasars changed slightly due to the deflection of the radio light from 3C 273. To a very high degree of accuracy, the astronomers' observations agreed with Einstein's general theory of relativity.

Perhaps the best way of understanding how the geometry of space-time affects the behavior of light rays and particles involves the use of embedding diagrams. As noted in earlier chapters, it is virtually impossible to visualize a warped four-dimensional space-time. To overcome this difficulty, theoretical physicists sometimes prefer to visualize processes in two dimensions and then generalize the results to four dimensions. Or, conversely, they are better able to understand what the equations are saying if they can suppress two of the four dimensions and look at the resulting curved surface. Figuratively speaking, this is done by slicing through warped space-time and examining the shape of the resulting surface. Such a procedure is like slicing through a cake to see how the layers of cake and icing are organized. A slice through space-time is called a *hypersurface,* and if the slice is taken perpendicular to the time axis, the hypersurface is said to be *spacelike.* A drawing of such a spacelike hypersurface is called an *embedding diagram.*

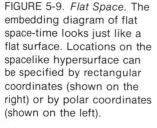

FIGURE 5-9. *Flat Space.* The embedding diagram of flat space-time looks just like a flat surface. Locations on the spacelike hypersurface can be specified by rectangular coordinates (shown on the right) or by polar coordinates (shown on the left).

To better understand the meaning of embedding diagrams, consider ordinary flat space-time such as is found in empty space far from any sources of gravity. The result of slicing through flat space-time is simply a flat two-dimensional hypersurface. This surface is flat in exactly the same sense that the floor or a table top is flat. A drawing of the surface, as shown in Figure 5-9, is an embedding diagram.

Now consider the warped space-time around the sun. The sun has not changed for billions of years and therefore the geometry of space-time has not changed. A spacelike hypersurface will look the same a billion years in the future as it did a billion years in the past.

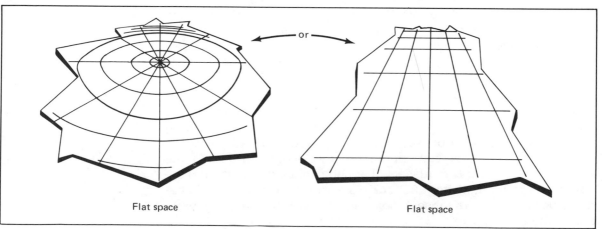

Flat space   Flat space

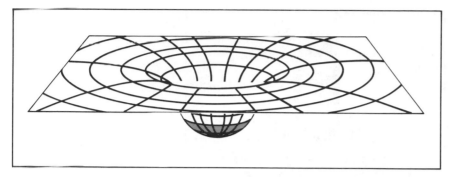

FIGURE 5-10. *Curved Space.* An embedding diagram vividly displays the curvature of space around the sun. The shaded area indicates the location of the sun. (Adapted from Misner, Thorne, and Wheeler.)

If this space-time is sliced through, however, the resulting hypersurface will *not* be flat because of the warping of the sun's gravitational field. An embedding diagram showing this warping is seen in Figure 5-10. The shaded area indicates the location of the sun. In essence, an embedding diagram tells us what gravity would do if we were living in two-dimensional space rather than four-dimensional space-time. An embedding diagram shows how gravity affects the curvature of space.

With the aid of an embedding diagram, the deflection of starlight (or radio waves from quasars) can be visualized easily. Since the hypersurface shown in Figure 5-10 is not flat, light rays traveling *on* this curved surface cannot move along straight lines. As seen in Figure 5-11, the geodesics followed by starlight are curved, and thus stars appear deflected from their usual positions.

While embedding diagrams are useful in visualizing the effects of

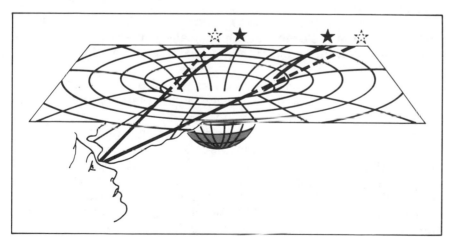

FIGURE 5-11. *The Deflection of Starlight.* The bending of light rays in general relativity can be intuitively appreciated with the aid of an embedding diagram. Light rays follow geodesics (that is, the shortest possible paths) *on* the hypersurface. Since the surface is curved, the paths are also curved. (Adapted from Misner, Thorne, and Wheeler.)

FIGURE 5-12. *The Slowing Down of Time.* Gravity causes time to slow down. Clocks on the ground floor of a building tick more slowly than clocks in the attic.

general relativity on the geometry of space, corresponding effects on time can be appreciated by examining the behavior of clocks. According to Einstein's theory, *gravity causes time to slow down.* The stronger the gravitational field, the greater the slowing down of time. For example, imagine two people in a house. One person is on the ground floor while the other person is in the attic, as shown in Figure 5-12. The person on the ground floor is closer to the earth and therefore in a slightly more intense gravitational field than the person in the attic. Upon comparing their wristwatches, they find that clocks on the ground floor measure time more slowly than clocks in the attic. This does *not* mean that a clock far from the earth would run very fast. In flat space-time far from any source of gravity, all clocks run at the same uniform rate. Clocks in a gravitational field run at a slower rate. Specifically, a clock on the earth's surface loses about one-billionth of a second each month compared to a clock out in space.

The gravitational slowing of time was the third effect predicted by Einstein as a test of general relativity. Unlike Mercury's precession and the deflection of light by the sun, this third effect is so small that scientists simply did not have clocks accurate enough to measure it. In the late 1950s, shortly after Einstein's death, the German physicist Rudolph Mössbauer discovered an important effect in nuclear physics. The *Mössbauer effect,* for which its discoverer was awarded the Nobel Prize, provides a method of using atomic nuclei as extremely sensitive clocks. Although this important discovery has numerous practical applications, in 1959 R. V. Pound and G. A.

78

Rebka at Harvard University realized that the Mössbauer effect could be used to test general relativity.

Anything that emits light may be thought of as a clock. When atoms emit light, they do so at specific wavelengths or frequencies, and time can be measured by measuring the frequency (in cycles per second, for example) of that light. Since gravity slows down time, light emitted by atoms in a gravitational field is *redshifted* to longer wavelengths or lower frequencies (that is, fewer cycles per second). Einstein's prediction of the slowing of time is therefore often called the *gravitational redshift.*

The light from atoms cannot be used to measure the gravitational redshift on earth because atoms do not emit light at frequencies precise enough to detect the minute slowing of time at the earth's surface. The Mössbauer effect, however, involves the emission of gamma rays by radioactive nuclei such as cobalt (^{60}Co) or iron (^{57}Fe). Using the discoveries of Mössbauer, we find that radioactive isotopes can emit gamma rays at incredibly precise frequencies. Pound and Rebka realized that the precision is high enough to detect the gravitational redshift here on earth.

The Pound-Rebka experiment was performed in the Jefferson Physical Laboratory at Harvard University. Gamma rays were emitted from radioactive cobalt (^{57}Co) in the basement. They traveled through holes drilled in the various floors to an absorber located in the penthouse 74 feet up. Upon measuring the absorbed gamma rays, Pound and Rebka found that the frequency was lower by exactly the amount predicted from Einstein's theory. This experiment was repeated in 1965 by R. V. Pound and J. L. Snider with the same result. The gravitational slowing of time had finally been detected (see Figure 5-13).

The publication of Einstein's paper "Zür allgemeinen Relativitätstheorie" late in 1915 had a profound effect on all science. For over two centuries Newtonian mechanics had appeared correct in virtually every problem involving gravitation. Yet now there was a new theory which worked even better and which required fundamental reorientation of our concepts of space and time. In spite of the initial excitement, scientists soon realized that the effects of general relativity were extremely difficult to detect. So everybody promptly went back to using the old-fashioned Newtonian theory. After all, the mathematics was a lot simpler than trying to solve the field equations and, at worst, thinking of gravity as a force always gave 99.99 percent of the correct answer. General relativity looked very uninteresting.

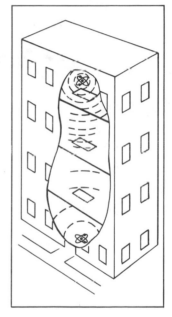

FIGURE 5-13. *The Gravitational Redshift.* Time slows down in a gravitational field. Therefore gamma rays emitted from radioactive nuclei in the basement of a building should have a lower frequency than corresponding gamma rays emitted by an identical source in the penthouse.

During the late 1960s, interest in the effects and predictions of general relativity was rekindled. This rebirth of relativity is due primarily to the fact that astrophysicists have an improved understanding of the life cycles of stars. As will be seen in the next two chapters, the death of a star can result in gravitational fields so intense that space-time folds in over itself and the star disappears from our universe. What is left is called a *black hole.* The curvature of space-time around a black hole is so great that there is a place where time stops completely! The gravitational redshift, instead of being a tiny effect, dominates everything.

The gravitational redshift effect is always a *slowing* of time. This is a direct result of the fact that gravity is always attractive. Scientists have never observed repulsive gravitation or antigravity. Yet in discussing rotating black holes, we will find that it is theoretically possible to travel *through* such black holes to regions of space-time where gravity is negative. In these regions of antigravity, time speeds up and clocks go fast. Antigravity is for people in a hurry!

6 STARS AND STELLAR EVOLUTION

As we look up into the star-filled night sky, the heavens seem permanent and unchanging. Even to the most careful observer, the patterns of stars that make up the constellations appear almost the same today as they did thousands of years ago. The brightest stars are still the brightest; the dimmest are still the dimmest. Yet, with a little thought one realizes that the apparent permanence of the heavens must be an illusion. We see the stars in the sky because they give off light. As they emit light, they use up energy. The depletion of their energy sources must result in changes inside these stars. Stars must evolve.

Imagine a small insect in the forest. Suppose that this insect is blessed with great intelligence, yet at the same time cursed with a very short life span. It will live for only 24 hours. As this insect looks around the forest it sees huge trees towering high above. It sees green shoots sprouting through the moist soil and occasional rotting logs scattered randomly over the ground. To this insect the forest seems eternal and unchanging. During its entire lifetime the insect will not make a single observation that would contradict this first impression. Yet by using its intellect, it comes on a fascinating idea. Maybe the forest does change. Perhaps the small green shoots grow to become trees. Perhaps the most ancient trees eventually fall to the ground, producing the rotting logs that enrich the soil for future generations of trees. Although the forest seems unchanging, this intelligent insect is able to discover the life cycle of the trees around it.

In order to discover the life cycles of stars, astronomers must begin by understanding what stars are. Looking into the sky, they see bright stars and dim stars, bluish stars and reddish stars. Astronomers immediately realize that first impressions can be very misleading. For example, if you see a bright star in the sky, you have no way of knowing how bright the star *really* is. It might be an extremely luminous star very far from the earth, or it might be a dim star that just happens to be nearby. The *apparent* brightnesses of stars do not tell astronomers anything of a fundamental nature about the intrinsic properties of stars. Astronomers would prefer to know the *absolute* brightnesses of stars. The absolute brightness of a star tells them how bright the star really is. It tells them how much energy the star is emitting into space.

The apparent brightness and absolute brightness of a star are related to each other by the distance to the star. To see why this is so, imagine looking at a street light on a dark night. Just by observing how bright the light appears to your eyes, you have no way of knowing how bright the light really is. It could be a 100-watt lamp that is nearby or a 500-watt lamp far away. But if the distance to the light is also known, it is then possible to calculate how bright the light really is. There is a very simple relationship between apparent brightness, absolute brightness, and distance. If the apparent brightness and distance are known, it is always possible to calculate the real or absolute brightness. This absolute brightness then tells you a fundamental property of the source of light. It tells you exactly how many watts the lamp (or star) is actually emitting.

Beginning in the mid-1800s, astronomers had finally developed the technique of parallax to such an extent that the distances to many stars could be measured. Parallax, as described in Chapter 1, is a straightforward but painstaking method of measuring the distances to the stars directly. The final result was that, from knowing the distances, astronomers easily calculated the absolute brightnesses of stars. They finally began learning how bright the stars really are.

A convenient way of expressing the absolute brightness or *luminosity* of a star is simply to say how many times brighter or dimmer the star is compared to the sun. Of course, the luminosity of the sun is set exactly equal to 1. The sun shines with the luminosity of "1 sun." Many stars emit only one-hundredth the amount of light given off by the sun. Their luminosities are approximately "$1/100$ suns." On the other hand, many stars emit thousands of times more light than the sun. For example, the bright, bluish star Rigel in the constellation of Orion has a luminosity of 50,000 suns. Rigel happens to be one of the intrinsically brightest stars known to astronomers.

In addition to knowing the real luminosities of stars, astronomers would also like to know how hot stars are, what they are made of, and how much matter they contain. Headway into some of these problems was made possible by a series of remarkable discoveries also beginning in the mid-1800s. Since the time of Isaac Newton, it has been known that white light passing through a glass prism is broken up into the colors of the rainbow. Such a rainbow of colors is called a *spectrum* (see Figure 6-1). In 1815, however, the German physicist Joseph Fraunhofer noticed that there are faint dark lines interspersed among the colors of the sun's spectrum. The true nature of these *spectral lines* remained a mystery until the early twentieth century, when it was discovered that they are formed by the various chemicals contained in the source of light. As a result of the work of great physicists like Max Planck and Niels Bohr, it was

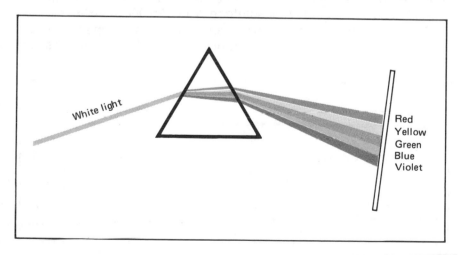

White light

Red
Yellow
Green
Blue
Violet

FIGURE 6-1. *A Spectrum.* When light is passed through a prism, the white light is broken up into the colors of the rainbow. This display is called a *spectrum,* and often contains tiny black lines formed by the chemicals in the source of light.

proved that spectral lines are caused by electrons changing orbits inside atoms. As an electron goes from one orbit to another, it absorbs or emits light of a specific wavelength. This absorption or emission of light produces patterns of spectral lines. Different chemicals are made up of different kinds of atoms and therefore produce different yet distinctive patterns of spectral lines. In other words, the chemicals in a source of light leave their fingerprints on that light in the form of patterns of spectral lines. By identifying spectral lines, the physicist or astronomer can discover the chemical composition of the source of light.

From examining the spectra of stars, astronomers finally had the necessary knowledge to discover what stars are made of. As a result of years of study, it is now known that stars consist almost entirely of hydrogen and helium. Hydrogen, the lightest element, comprises 50 to 80 percent of the matter in stars. Hydrogen and helium together account for 96 to 99 percent of the mass of most stars. This leaves less than 4 percent for all the heavy elements together. The most abundant of these heavy elements are oxygen, nitrogen, carbon, neon, magnesium, argon, chlorine, silicon, sulfur, and iron.

Even though all stars have roughly the same chemical composition, the spectra of stars vary widely. For example, Figure 6-2 shows spectra of three typical stars. They all are made out of essentially the same chemicals, yet their spectra are dominated by very different patterns of spectral lines because these three stars have very different temperatures. The temperature of a star's atmosphere has a profound effect on precisely which spectral lines of which chemicals show up strongest. Imagine a very hot star whose surface temperature is 25,000 degrees above absolute zero (i.e., 25,000 °K). Gases in the star's atmosphere are so hot that many atoms are stripped of outer electrons. Such atoms cannot produce spectral lines in the visible

FIGURE 6-2. *Stellar Spectra.* The three stars α Canis Majoris, τ Scorpii, and β Pegasi have almost exactly the same chemical composition. Their spectra look very different because the surface temperatures of the three stars are very different. (Hale Observatories)

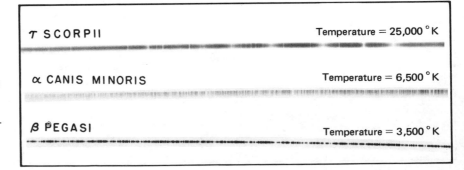

τ SCORPII   Temperature = 25,000 °K

α CANIS MINORIS   Temperature = 6,500 °K

β PEGASI   Temperature = 3,500 °K

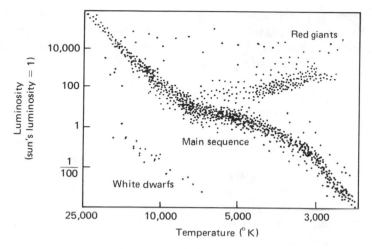

FIGURE 6-3. *The Hertzsprung-Russell Diagram.* The luminosities and temperatures of stars are displayed most conveniently in the form of a graph. Most stars are located in one of three basic regions: the main sequence, red giants, or white dwarfs.

spectrum. Specifically, only helium atoms can manage to hold on to all their electrons at such temperatures. Consequently, stars whose spectra are dominated by helium lines must have surface temperatures of approximately 25,000 °K. As another example, consider a cool star whose surface temperature is only 3,000 °K. At such low temperatures, atoms can combine to form molecules. The spectra of such stars are dominated by lines due to molecules such as titanium oxide, even though titanium is a rare element. Finally, the sun's spectrum shows many spectral lines of metals such as calcium, magnesium, and silicon. This tells astronomers that the surface temperature of the sun must be about 6,000 °K, because at such a temperature conditions are ideal (not too hot, not too cool) for the formation of spectral lines by metals.

By way of summary, the study of stellar spectra tells astronomers what stars are made of. But, perhaps more importantly, by examining the spectrum of a star, it is possible to deduce the surface temperature of that star. In this way astronomers learn how hot stars are.

Knowing the real luminosities and surface temperatures of the stars, astronomers can display this information in a very useful fashion. Shortly before the beginning of World War I, the Danish astronomer Einar Hertzsprung and the American astronomer Henry Norris Russell independently discovered that an interesting graph was obtained by plotting luminosities versus temperatures of stars. As shown in Figure 6-3, the luminosities of stars are measured along the vertical axis. The surface temperatures of stars are measured along the horizontal axis. Every star in the sky for which the luminosity and temperature are known can be represented by a point on the resulting graph. For example, the sun, whose luminosity is 1 and whose surface temperature is about 6,000 °K, is represented by a dot near the middle of the diagram. This graph is called the *Hertzsprung-Russell* diagram in honor of the astronomers who invented it.

Notice that the dots representing stars in the sky are *not* scattered randomly all over the Hertzsprung-Russell diagram. Instead, the dots are generally grouped in three basic regions. Most stars seen in the sky are located along the *main sequence.* The main sequence runs diagonally across the Hertzsprung-Russell diagram from the hot, bright stars in the upper left to the cool, dim stars in the lower right. The dot representing the sun is near the middle of the main sequence. It is therefore said that the sun is a *main-sequence star.*

In addition to the main sequence, there is a second major grouping of stars in the upper right-hand corner of the Hertzsprung-Russell diagram. These stars are both bright and cool. They emit thousands of times more light than the sun, yet have surface temperatures between 3,000 °K and 4,000 °K. In addition, these stars are enormous. If one of these stars were placed at the center of the solar system, its surface would extend beyond the orbit of the earth. Their diameters are typically several hundred million miles. Finally, since these stars are so cool, they emit primarily reddish light. They are therefore called *red giants.*

Almost every reddish star you can see in the sky is a red giant. Betelgeuse in Orion, Antares in Scorpius, and Aldebaran in Taurus are good examples. All other stars visible to your naked eyes are main-sequence stars.

With a good telescope it is possible to discover another class of stars that are neither red giants nor main-sequence stars. This third class consists of stars that are very hot and very dim. Their surface temperatures are typically between 10,000 °K and 20,000 °K, and they emit only one-hundredth of the light given off by the sun. Dots representing such stars therefore appear in the lower left-hand corner of the Hertzsprung-Russell diagram. Stars that are very hot emit primarily bluish-white light. In addition, these hot, dim stars are very small. They are typically about the same size as the earth, having diameters of only 10,000 miles, and are called *white dwarfs.*

The importance of the Hertzsprung-Russell diagram cannot be overemphasized. In many respects, it can easily be called *the* most important graph in all astronomy. For some reason, most stars are *either* main-sequence stars *or* red giants *or* white dwarfs. Of course,

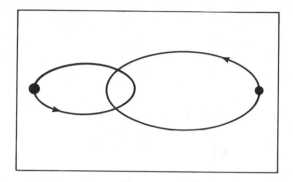

FIGURE 6-4. *A Binary Star.* Many stars seen in the sky actually consist of two stars very close together which are in orbit about their common center.

there are a few exceptions but, nevertheless, most stars spend billions of years as one of these basic types of stars.

At the beginning of this chapter it was noted that stars must evolve. This means that during its entire life cycle, a star changes its luminosity and surface temperature. In other words, the dot representing a given star must move around on the Hertzsprung-Russell diagram. Understanding how stars move around on the Hertzsprung-Russell diagram is therefore the same as knowing how stars are born, what they look like when they mature, and what happens to them as they die.

Before astronomers can begin to tackle the problem of trying to understand the life cycles of stars, they need one more important piece of information. They need to know exactly how much matter is contained in stars; they need to know stellar masses.

Surprising as it may seem, about half the stars seen in the sky are not individual stars like our sun. Instead they consist of *two* stars revolving about their common center just as the earth and the moon revolve about each other. See Figure 6-4. Such stars are called *double stars* or *binary stars.* Binary stars are very important to astronomers because it is possible to deduce the masses of the stars from observing the motions of stars in a binary. From seeing how the two stars revolve about their common center, astronomers can use Newtonian mechanics to discover how massive the stars must be. In this way astronomers learn how much matter is contained in stars.

The results of measuring the masses of many binary stars are most easily displayed in the form of a graph, as shown in Figure 6-5. It is found that the dimmest stars are the lowest-mass stars. Such stars typically contain only a tenth of the amount of matter in the sun. On the other hand, the most luminous stars are the most mas-

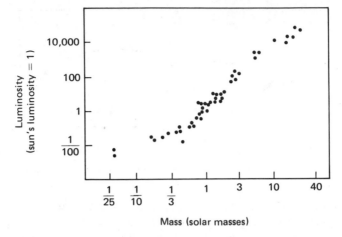

FIGURE 6-5. *The Mass-Luminosity Relation.* The masses and luminosities of main-sequence stars are correlated as shown in this graph. Dim stars have the lowest masses (¹/₁₀ solar mass or even less), while the brightest stars are the most massive (up to 50 solar masses).

sive. Stars of 40 and 50 solar masses are known to astronomers. This correlation between the mass and luminosity of main-sequence stars is called the *mass-luminosity relation.*

Knowing the luminosities, temperatures, and masses of many stars, astronomers now turn to the astrophysicists to find out what is really going on. Why are stars arranged in three major groups on the Hertzsprung-Russell diagram? Why are the most massive stars also the most luminous? How are main-sequence stars related to red giants? Are red giants somehow related to white dwarfs? Or to something else? The astrophysicists then proceed to take the laws of physics, mathematics, and the astronomers' data and program it all into huge computers. In a matter of minutes (or sometimes hours) these computers reproduce mathematically what it takes nature billions of years to do in the skies. The final result is the fascinating story of *stellar evolution.*

As astronomers look up into the heavens, they often find large clouds of gas. An excellent example is the Orion nebula (sometimes called M 42), which is just barely visible to the naked eye as the middle "star" in the sword of Orion. A beautiful photograph of this nebula is shown in Figure 6-6. Notice that there are some dark regions in the nebula. These are not "holes" in the nebula, as was thought back in the nineteenth century. Instead, they are cool, dark clouds of dust obscuring our view of the more luminous gases beyond.

Think carefully about one of these cool, dark clouds of gas and dust. As you might expect, it is not perfectly homogeneous but con-

FIGURE 6-6. *The Orion Nebula.* Large, cool clouds of gas, such as this nebula in Orion, contain the birthplaces of stars. (Lick Observatory)

tains lumps where the gas is slightly denser than the surrounding regions of the cloud. Since one of these lumps contains more matter than the surrounding regions, it will have a slightly higher gravitational field. The lump will therefore attract nearby matter. In doing so, the lump becomes more massive; it then has a stronger gravitational field which attracts still more matter. By this method of *accretion,* the lump grows in size and mass until it finally contains a huge amount of matter many times the mass of the sun spread over a volume many times the size of the solar system.

Detailed calculations by astrophysicists reveal that such a *protostar* is unstable. There is simply nothing to hold up the crushing inward weight of trillions upon trillions of tons of gas. The protostar therefore begins to contract. As all the matter in this huge ball of gas is squeezed into a smaller and smaller volume, the pressures

FIGURE 6-7. *The Pleiades*. A cluster of very young stars. Only a billion years ago, thermonuclear fires were ignited at the centers of these stars. (Lick Observatory)

and densities inside the protostar begin to rise dramatically. When you rub your hands together, your palms get warm. For much the same reason, as the protostar contracts, the temperature near its center gets higher and higher. Finally, when the temperature at the center reaches about ten million degrees, the nuclei of hydrogen atoms collide with such violence that they *fuse* together to form the nuclei of helium atoms. This *thermonuclear reaction,* whereby hydrogen is converted into helium, releases an enormous amount of energy. This is the same process that occurs in hydrogen bombs. The violent outpouring of energy is sufficient to stop the contraction. A star is born!

During the process of contraction, the point representing the protostar is moving very rapidly across the Hertzsprung-Russell diagram because conditions at the protostar's surface are changing rapidly. Initially, as the protostar decreases in size, its luminosity decreases. Later on, just before thermonuclear reactions are ignited, the surface temperature of the star rises rapidly. According to the astrophysicists' calculations, the point on the Hertzsprung-Russell diagram stops moving when hydrogen burning starts at the star's core. The point comes to rest *on* the main sequence. See Figure 6-7 for an excellent example of a young cluster of stars.

In this way astrophysicists have been able to discover the true meaning of the main sequence. Every star on the main sequence has hydrogen burning occurring at its center. In massive stars, hydrogen burning occurs at a furious rate. The most massive stars are therefore the brightest stars. In low-mass stars, hydrogen burning occurs at a much slower rate. The least massive stars are therefore also the dimmest stars.

In the sun, a typical garden-variety main-sequence star, 600 million tons of hydrogen are converted into helium each second. Although this may sound like an astonishingly rapid rate, there is so much hydrogen gas at the sun's core that the sun can sustain this rate for a total of ten billion years. For ten billion years, the dot representing the sun on the Hertzsprung-Russell diagram sits smack in the middle of the main sequence. During this entire time, the sun—or any similar star—changes very little. The sun is now about five billion years old, and we have about five billion years left.

Eventually all the hydrogen at the core of a main-sequence star is used up. The depletion of hydrogen signals the onset of some drastic changes. Recall that the onset of hydrogen burning was the process responsible for stopping the initial contraction of the protostar. Therefore, as soon as the hydrogen burning shuts off, the star's core begins to contract; there is simply nothing to hold it up. As the core shrinks, pressures, densities, and temperatures again rise dramatically. Finally, when the temperature at the center of the star reaches 100 million degrees, the nuclei of helium atoms (left over from the hydrogen-burning stage) are violently fused together to form carbon. This ignition of helium burning at the star's core produces a tremendous outpouring of energy. In addition, the release of energy inside the star as the core contracts pushes the surface of the star out into space in all directions. As the star expands, the gases in its atmosphere cool down to between 3,000 °K and 4,000 °K. The result is an enormous star a quarter of a billion miles in diameter with a low surface temperature: a red giant.

About five billion years from now, all the hydrogen at the sun's core will have been used up. Its core will collapse while its surface expands and helium burning is ignited. In a fairly short period of time (less than a billion years) the phenomenal expansion of the sun will swallow the earth and our planet will be vaporized.

Just as hydrogen was depleted, all the helium at the center of a red giant is eventually used up. This results in still further collapse

FIGURE 6-8. *A Nova.* In the spring of 1934, a star in the constellation of Hercules blew up. These are the kinds of violent events that occur as stars in a binary system near the end of their life-cycles. (Lick Observatory)

of the star's core and, for stars considerably more massive than our sun, the ignition of still more exotic thermonuclear reactions such as carbon burning, oxygen burning, and silicon burning. It is during these processes that all the heavier elements are created in the most massive stars.

Although the precise details are not currently understood, it is believed that a star in such a late stage of evolution becomes extremely unstable (Figure 6-8). For example, such a star can pulsate in size and luminosity.

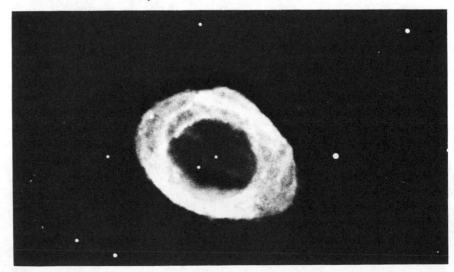

FIGURE 6-9. *A Planetary Nebula.* The Ring nebula in Lyra is the result of a dying star ejecting its outer atmosphere into space. (Lick Observatory)

92

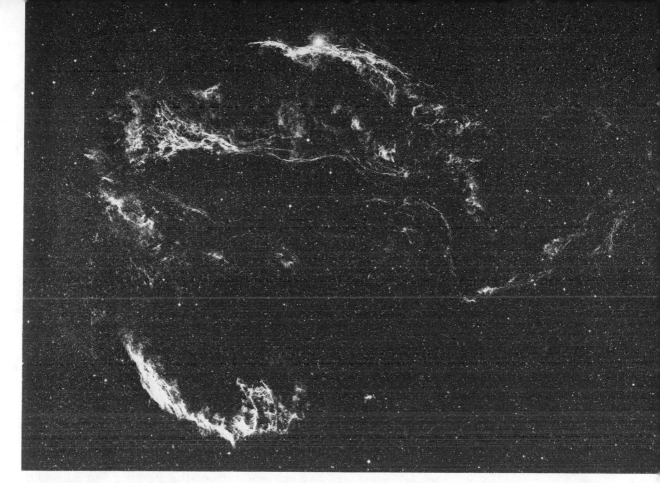

Finally these instabilities inside a massive star nearing the end of its life can become so great that the star ends its existence in a spectacular detonation. These explosions are sometimes so colossal that the dying star momentarily outshines the entire galaxy in which it resides. Such an exploding star is called a *supernova*.

In going through its death throes, a dying star can eject great quantities of matter into space. The gases sometimes appear in the form of *planetary nebulae* such as the Ring nebula in the constellation of Lyra, shown in Figure 6-9. More violent supernova explosions, which can occur during the death of a very massive star, leave behind *supernova remnants* such as the Veil nebula in Cygnus, shown in Figure 6-10. Often more than 25 percent of the mass of a star is ejected into interstellar space in producing these various nebulae.

The death of a star leaves behind a burned-out stellar core. In the case of low-mass stars like the sun, this core shrinks in size until certain forces (discussed in the next chapter) resist further collapse. At this stage it is very hot and very small. It is a white dwarf.

As a result of a lengthy series of elaborate calculations since the

FIGURE 6-10. *A Supernova Remnant.* Roughly 50,000 years ago, a dying star in the constellation of Cygnus underwent a supernova explosion that produced this beautiful object appropriately called the Veil nebula. (Hale Observatories)

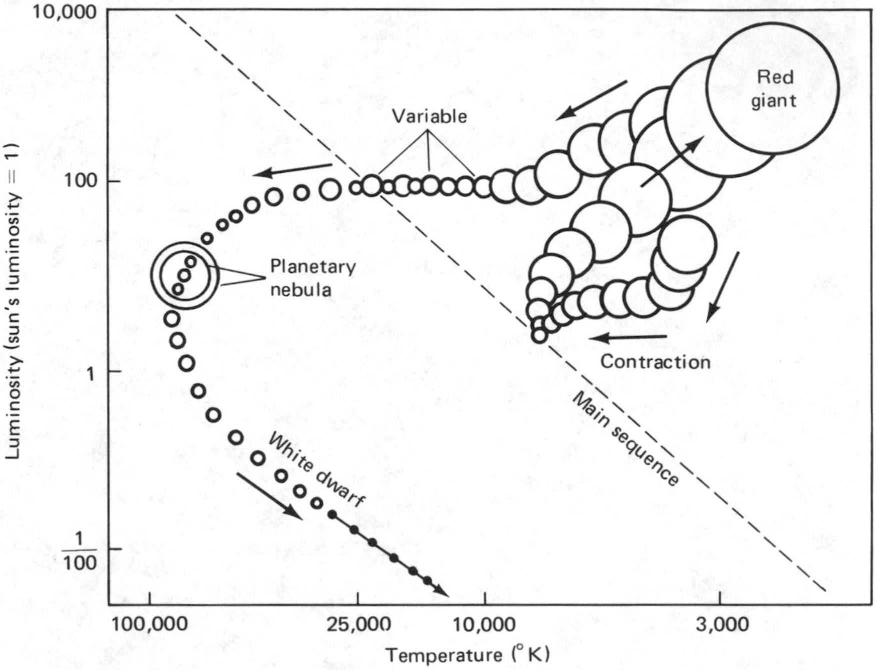

FIGURE 6-11. *The Evolution of a Star.* The entire life cycle of a star like the sun can be represented as a single dot moving around on the Hertzsprung-Russell diagram. The dot spends very long periods of time first on the main sequence and then in the red-giant region. Finally, it ends as a white dwarf. Transition regions are traversed very rapidly.

early 1960s, the entire life cycle of a star like the sun can be expressed in terms of a point (representing the star) moving around on the Hertzsprung-Russell diagram. As shown in Figure 6-11, the initial contraction of the protostar results in a rapid drop in luminosity as the protostar decreases in size. This is followed by a rise in surface temperature as the star's atmosphere heats up. When hydrogen burning commences at the star's center, the point settles down on the main sequence and remains there for almost ten billion years. The transition to the red-giant region is again very rapid. But once helium burning is ignited, the point remains in the upper right-hand corner of the Hertzsprung-Russell diagram for several hundred million years. With the onset of instabilities, the point again begins moving rapidly across the Hertzsprung-Russell diagram. Although the evolutionary tracks are not precisely known (the star is changing so fast that computers cannot handle the calculations) the star ends up as a white dwarf. White dwarfs are *dead stars.* They just sit out in space and cool off just as a hot cup of coffee left on the kitchen table cools off. As white dwarfs cool, they gradually get dimmer and dimmer. The point representing a white dwarf slowly moves along *cooling curves,* which are sloped to the lower right in the diagram.

94

There are several important facts about stellar evolution that should be emphasized. First of all, the most massive stars on the main sequence are also the most luminous. They are bright because they are burning hydrogen at a furious rate. In spite of their large mass and thus enormous reserves of fuel, all the hydrogen in the cores of such stars is depleted very rapidly. In other words, the most massive stars evolve the most rapidly.

Second, from studying planetary nebulae and supernova remnants we are led to believe that the most massive stars can eject some, but perhaps not all, of their matter into space. The rest remains behind to form a stellar corpse.

And finally, as discussed in the next chapter, astrophysicists are absolutely certain that there is a strict upper limit to the mass of a white dwarf. White dwarfs must have masses *less* than 1¼ solar masses.

So there seems to be a severe problem. Imagine a massive star like Rigel in Orion. A star like Rigel has a mass of 40 suns and therefore evolves very rapidly. But in order to become a white dwarf, it must eject almost 39 solar masses of matter into space. This appears to be an excessively large percentage from what astronomers know about planetary nebulae and supernova remnants.

Up until the mid-1960s it was generally believed that somehow even the most massive stars manage to blow off enough matter to get below the critical limit for the masses of white dwarfs. But by the late 1960s, certain remarkable discoveries were made by radio astronomers which shed serious doubt on this majority opinion. As a result of these doubts, astrophysicists in the early 1970s began seriously considering the possible existence of some of the most bizarre objects ever imagined by the rational human mind: black holes.

7 WHITE DWARFS, PULSARS, NEUTRON STARS

One of the most important breakthroughs in understanding the nature of light occurred in the mid-1800s as a result of Maxwell's formulation of electromagnetic theory. As discussed in Chapter 2, Maxwell's theory leads directly to a fundamental wave equation that completely describes all the wave properties of light. Surprisingly, this wave equation places no restrictions on the possible wavelengths of light. Yet visible light, to which our eyes respond, is confined to a small specific range of wavelengths, from about 0.000016 inch for violet light to 0.000028 inch for red light. In other words, from Maxwell's theory there should be other types of electromagnetic radiation at much longer and much shorter wavelengths outside the limits of visible light. In this sense, the electromagnetic theory of light actually predicted the existence of x-rays, gamma rays, radio waves, and ultraviolet and infrared radiation.

Astronomers had good reason to be disquieted by this discovery. Virtually everything known about the universe had come from optical observations of visible light. But visible light constitutes a very tiny fraction of the entire electromagnetic spectrum (see Figure 2-5). Perhaps stars and galaxies also emit other types of radiation, such as x-rays and radio waves, in addition to visible light. Since the human eye cannot see any of these radiations, astronomers became painfully aware that they were missing a lot of information.

The difficulties of observing the heavens at wavelengths far removed from visible light are twofold. First of all, until very recently, scientists simply did not know how to construct instruments capable of detecting exotic forms of electromagnetic radiation. The human eye or a piece of photographic film is of no help at all. Second, the earth's atmosphere is opaque to much of the electromagnetic spectrum. X-rays and ultraviolet radiation simply do not penetrate the blanket of air surrounding the earth's surface.

The opacity of the earth's atmosphere is at the same time a blessing and a curse. Because the atmosphere shields us from many of the deadly radiations from outer space, life has been able to evolve on our planet's surface. Observations of the sky at certain wavelengths, however, simply cannot be made from the ground since certain types of radiation do not penetrate the earth's protective blanket of air. Trying to see the x-ray or ultraviolet sky from the earth's surface is like trying to look through a brick wall. The atmosphere shields us but also keeps us in ignorance.

From this historical perspective we can appreciate the implications of the work of Karl G. Jansky, who, in 1931, detected radio waves from outer space. Jansky was experimenting with antennas for Bell Telephone Laboratories when he noticed that his instruments were responding to radio waves from sources beyond the earth. Radio waves as well as visible light penetrate the earth's atmosphere.

World War II severely interfered with a wide range of scientific endeavors, including any substantial developments in the new field of radio astronomy. But as a result of major advances in electronics and electrical engineering associated with the war effort, astronomers in the late 1940s realized that they had many of the necessary tools to construct radio telescopes. During the 1950s, numerous radio telescopes were built around the world and scientists had their first opportunity to see what the radio sky really looks like.

The invention of radio telescopes (Figure 7-1) gave people a new set of eyes with which to view the universe. For the first time, scien-

FIGURE 7-1. *A Radio Telescope*. A radio telescope consists of a huge metal dish that focuses, collects, and magnifies the radio waves from outer space. (NASA, Jet Propulsion Laboratory)

tists had the ability to observe the universe at wavelengths far removed from the visible part of the electromagnetic spectrum.

Most of the 1950s were spent in making accurate records of the appearance of the radio sky (Figure 7-2). Radio astronomers spent many years making catalogs of the radio sources they discovered. Thousands of objects are now listed in such catalogs. By the 1960s, however, radio astronomers finally had the time to devote to detailed observations of many of the fascinating radio sources they had discovered.

In the fall of 1967, a team of radio astronomers headed by Antony Hewish at Cambridge University was making some observations when they noticed that their radio telescope was detecting very unusual signals. They were observing short pulses of radio noise arriving at approximately one-second intervals (see Figure 7-3).

It was initially supposed that they had detected signals from a Soviet satellite launched in secret, but the possibility that these signals were coming from an artificial spacecraft was soon discarded. A sat-

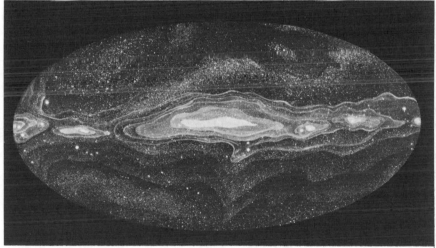

FIGURE 7 2. *The Visible and Radio Skies.* The upper drawing shows the entire visible sky. The drawing is oriented so that the Milky Way runs horizontally across the diagram. The lower drawing shows the corresponding appearance of the radio sky. (Lund Observatory; Griffith Observatory)

ellite inside the solar system would appear to move slowly with respect to the background stars as it orbits the sun. The mysterious pulsating radio source, however, remained fixed among the stars, indicating that it must be extremely remote. It must be out among the stars.

More detailed observations resulted in the surprising discovery that this pulsating radio source is extremely precise and regular. Although some pulses are weak and some are strong, the arrival time of the pulses is incredibly regular. Indeed, nothing had ever been

99

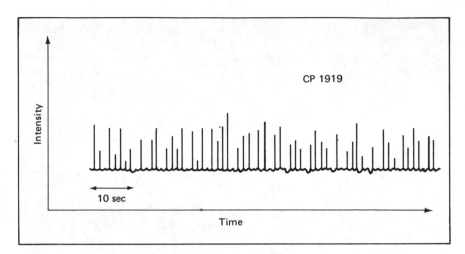

CP 1919

Intensity

10 sec

Time

FIGURE 7-3. *A Typical Pulsar Recording*. Radio bursts from pulsars are sometimes strong and sometimes weak. But the timing of the pulses is incredibly precise.

observed in nature which could rival the precision and regularity of these pulses. This observation prompted a host of ideas that we had succeeded in detecting signals from alien beings. The *LGM* (little green men) *Theory* hypothesized that the newly discovered radio source might be a navigational beacon for flying saucers.

By the spring of 1968, three additional pulsating radio sources, or *pulsars,* had been discovered. The first pulsar is called CP 1919; the remaining three are known as CP 0834, CP 0950, and CP 1133. (The letters CP stand for "Cambridge pulsar" and the numbers denote an approximate position in the sky.) In all cases, the periods of pulsation are extremely regular. Today, nearly 100 pulsars have been discovered by radio astronomers. Their periods range from $1/30$ of a second for the fastest to slightly over 3 seconds for the slowest.

Rather than appealing to little green men as an explanation for their discoveries, scientists prefer to examine what they know about the universe and see if they can understand their observations within the framework of the laws of nature. Thus, in the spring of 1968, astrophysicists around the world began a detailed reexamination of what they knew about the stars in order to find a rational and natural explanation of pulsars.

Recall from the previous chapter that thermonuclear reactions at the center of a star produce an outpouring of energy at enormous temperatures which is sufficient to hold up the outer layers of the star. Near the end of its life, when all the possible nuclear fuels have been exhausted, the star ends its existence in one of a variety of ways. Low-mass stars simply contract to become white dwarfs, perhaps shedding some of their matter. Stars of moderate mass can also become white dwarfs by shedding a greater percentage of their matter, perhaps forming a planetary nebula in the process. The most massive stars can end their lives in spectacular detonations called supernovae during which a large percentage of the dying star's matter can be violently ejected into space.

The nature of white dwarfs as "dead" stars has been well understood since the pioneering theoretical work of S. Chandrasekhar in the early 1930s. The thermonuclear fires that support ordinary stars cannot be responsible for holding up the outer layers of a white dwarf simply because all the fuel has been used up. To understand what does hold up a white dwarf, consider the matter inside the collapsing core of a dying star. As the star shrinks in size, the pressures and densities become so great that all the atoms are completely torn apart. This leaves the nuclei floating around in a sea of free electrons. Electrons are rotating or spinning and an important law of physics called the *Pauli exclusion principle* always applies to their behavior. In essence, this law says that no two electrons can occupy the same place with the same speed and spin at the same time. As the dying star collapses further and further, the electrons are squeezed together to such an extent that finally all possible locations and speeds for the electrons are occupied. As soon as this occurs, all the electrons act together powerfully to resist any further contraction of the dying star. In this way, the electrons produce a *degenerate electron pressure,* which supports the white dwarf against any further collapse.

White dwarfs have been known to astronomers for many years. These stars are so common that, until recently, it was generally believed that all dying stars become white dwarfs. An excellent example of one of these dead stars is shown in Figure 7-4 as a companion to Sirius in the constellation of Canis Major.

As a result of detailed calculations concerning the structure of white dwarfs, Chandrasekhar discovered that there is a strict upper limit to the mass of a white dwarf. Degenerate electron pressure can support a dead star *only* if the mass of the star is *less* than about 1¼ solar masses. If the mass of the dying star is significantly greater than 1¼ suns, the powerful forces between the electrons are insufficient to hold up the crushing weight of the star. This critical limit of nearly 1¼ solar masses is known as the *Chandrasekhar limit.*

Since white dwarfs are so common and since no other kinds of "dead" stars were known to astronomers, it was believed that all dying stars somehow manage to eject enough matter to get below

FIGURE 7-4. *A White Dwarf.* Sirius, the brightest-appearing star in the night sky, has a white dwarf companion. White dwarfs, such as the one shown here alongside Sirius, are hot, dim, and very small. (Courtesy of R. B. Minton.)

the Chandrasekhar limit. This majority opinion, however, did not stop some astrophysicists from hypothesizing what might happen to a dying star with a mass larger than 1¼ solar masses.

In 1934, W. Baade and F. Zwicky envisioned what might happen to a 1½- or 2-solar-mass dead star. Since degenerate electron pressure is not strong enough to halt the contraction, the star simply gets smaller and smaller. The pressure and densities rise until the electrons are squeezed *into* the nuclei of the atoms out of which the star is made. When this happens, the negatively charged electrons combine with the positively charged protons to produce copious neutrons. Finally, when the entire star has been almost completely converted to neutrons, the Pauli exclusion principle again begins to play an important role. The forces between the neutrons give rise to a *degenerate neutron pressure.* This more powerful pressure can halt the contraction, resulting in a new type of stellar corpse: a *neutron star.*

Five years later, in 1939, J. R. Oppenheimer and G. Volkoff presented numerous calculations attesting to the feasibility of these ideas. Yet, since no one had ever seen a neutron star, these prophetic ideas were largely ignored. Actually, astronomers simply did not know what to look for.

102

FIGURE 7-5. *The Crab Nebula.* The pulsar NP 0532 is located at the center of this supernova remnant. (Lick Observatory)

In the year A.D. 1054, ancient Chinese astronomers noticed that a "guest star" had appeared in the sky in the constellation of Taurus, the bull. This new star became so bright that it could be seen easily during broad daylight. It then faded completely from view.

When modern astronomers turn their telescopes to the location of the "guest star" given in ancient records, they find the beautiful Crab nebula, shown in Figure 7-5. The Crab nebula is an excellent example of a supernova remnant, and the ancient Chinese astronomers were fortunate enough to observe a dying star blowing off its atmosphere.

In the fall of 1968, astronomers were amazed to find that a pulsar is located exactly in the middle of the Crab nebula. This pulsar, known as NP 0532, is the fastest of all known pulsars. Bursts of radio radiation are detected thirty times a second. This discovery led astronomers to suspect that dying stars might be associated with pulsars. After some straightforward calculations, it became obvious that white dwarfs could not produce thirty bursts of radio noise each second. It was time to resurrect the ideas of Baade, Zwicky, Oppenheimer, and Volkoff.

All stars rotate and all stars probably have magnetic fields. Under ordinary circumstances, these two properties are relatively unimpor-

103

FIGURE 7-6. *The Conservation of Angular Momentum.* As an ice skater doing a pirouette pulls in her arms, her rate of rotation speeds up.

tant. For example, the sun rotates once about its axis every month. In addition, the sun's magnetic field is fairly weak. The overall strength of the sun's magnetic field is roughly the same intensity as the earth's magnetic field. However, if the sun or any similar star were to collapse down to the size of a neutron star, these two properties would become extremely important. To see why this is so, imagine an ice skater doing a pirouette on the ice. As shown schematically in Figure 7-6, as she pulls in her arms, her rate of rotation speeds up. This is a direct result of a basic law of physics known as the *conservation of angular momentum.* Similarly, if a large star, such as the sun, shrinks down to a small size, its rate of rotation will increase dramatically. For this reason, astronomers believe that neutron stars rotate very rapidly, perhaps sometimes faster than once a second.

When a star is very large, its magnetic field is spread over millions of square miles of the star's surface. The intensity of the magnetic field at any location is therefore fairly weak. However, as a star dies, it shrinks in size. The magnetic field that was originally spread over a huge area is then confined to only a few hundred square miles. As the magnetic field's area is compressed, its intensity rises dramatically. If a star like the sun were to collapse down to the size of a neutron star, the strength of the magnetic field would increase by a factor of about a billion!

By way of summary, astronomers thinking about the probable structure of a neutron star have good reason to believe that the neutron star would be rapidly rotating and have an intense magnetic field. But they further realize that the axis of rotation is probably *not* parallel to the magnetic axis of the neutron star. After all, the earth's magnetic axis is tilted with respect to the earth's axis of rotation; a compass does *not* point to "true north."

Based on these considerations, it is possible to draw a diagram (Figure 7-7) illustrating the basic properties of a neutron star. The neutron star itself is very small, perhaps only 10 to 15 miles in diameter. Its rate of rotation is very high, and the intense magnetic field is inclined to the axis of rotation.

The important point to realize is that although the interior of the neutron star consists almost completely of neutrons, there are still plenty of charged particles (protons and electrons) on the star's surface. As these charged particles encounter the intense magnetic fields at the magnetic north and south poles, the particles are accelerated and therefore emit great quantities of radiation. In other

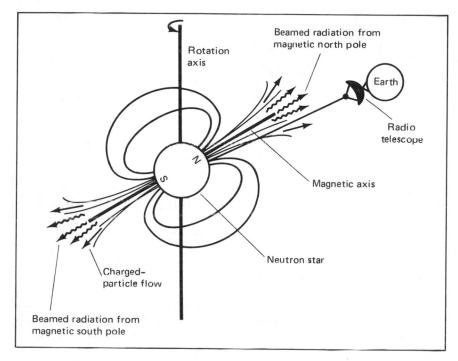

Rotation axis

Beamed radiation from magnetic north pole

Earth

Radio telescope

Magnetic axis

Neutron star

Charged-particle flow

Beamed radiation from magnetic south pole

FIGURE 7-7. *Details of a Neutron Star.* Astronomers believe that neutron stars are rapidly rotating and have intense magnetic fields. Radiation streaming out of the north and south magnetic poles can account for the properties of pulsars.

words, due to the interaction of the charged particles and the magnetic field, powerful beams of radiation should be emanating from regions near the north and south magnetic poles. But since the entire neutron star is rotating, these two beams of radiation will sweep around the sky. If the earth happens to be located in the path of one of these beams, radio astronomers will detect a pulse of radiation every time the beam passes in front of their line of sight. In this sense, the neutron star acts like the beam in a lighthouse. Every time the beam of light swings past the line of sight, a burst of radiation is observed.

All the known properties of pulsars can be explained in terms of this *oblique-rotator* model of neutron stars. Indeed, the theory fits the observations so well that few astronomers still doubt the existence of neutron stars. That which was once thought to be pure fantasy must actually exist in nature.

In thinking about rapidly rotating neutron stars being responsible for pulsars, one might wonder why only *radio* bursts are observed. Perhaps visible pulses might also be detected. This idea also oc-

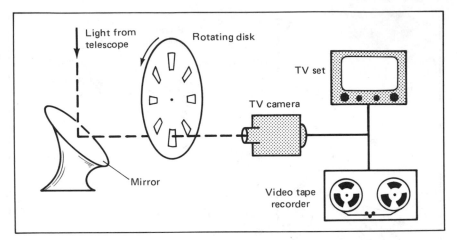

FIGURE 7-8. *TV Observations of NP 0532*. Visible pulses from the pulsar at the center of the Crab nebula were detected with the aid of a television camera and a rotating disk with slits that "chop" the incoming beam at various intervals.

curred to a team of astronomers at Kitt Peak Observatory in the fall of 1968. They trained their telescope on the stars at the center of the Crab nebula and recorded the images with a television camera, as shown in Figure 7-8. In between the telescope and the television camera they placed a rotating disk in which slits had been cut. By rotating the disk at various speeds, they could interrupt the incoming beam at various intervals. One of the stars was actually blinking on and off 30 times each second in *visible* light. However, in spite of exhaustive searches, only the Crab pulsar (NP 0532) has been observed flashing pulses of visible light (see Figure 7-9).

As the neutron-star explanation of pulsars gradually gained acceptance in the scientific community, astrophysicists began performing detailed calculations concerning the nature of "dead" stars. The laws of physics and extensive mathematics are programmed into a computer and the astrophysicist asks the computer to calculate specific details of the structure of white dwarfs or neutron stars. For example, Figure 7-10 shows the relationship between the mass and density of dead stars. Only two portions of the curve produced by the computer correspond to stable stars that can exist in nature. From this graph it is seen that the density at the center of a white dwarf is about 1,000 tons per cubic inch. At the center of a neutron star, however, the density is so high that one cubic inch contains about 40 billion tons of matter.

In addition to telling us the relationship between the mass and density inside a dead star, Figure 7-10 contains an important surprise. Notice that the Chandrasekhar limit stands out prominently. No white dwarf can exist with a mass greater than about 1¼ suns.

FIGURE 7-9. *The Pulsar NP 0532*. These two television pictures of the stars at the center of the Crab nebula show NP 0532 pulsating on and off in visible light. (Lick Observatory)

Degenerate electron pressure is simply not strong enough to support more than 1¼ solar masses. But there is *also* an upper limit to the mass of a neutron star. No neutron stars can exist with a mass greater than about 2¼ suns. Above this critical limit, degenerate neutron pressure is insufficient to hold up a dead star.

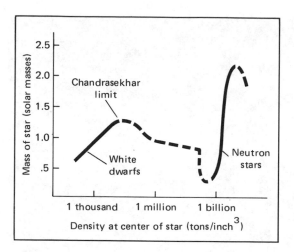

FIGURE 7-10. *Mass versus Density of Dead Stars.* The density of matter inside a white dwarf is about 1,000 tons per cubic inch. In a neutron star, the density is about 40 billion tons per cubic inch.

Figure 7-11 shows the relationship between the masses and diameters of dead stars. Again, it is seen that there are upper limits to the masses of white dwarfs and neutron stars. It might be thought that if rapid rotation were included these upper limits might be greatly increased. An atom in a rotating star experiences the inward force of gravity *and* an outward centrifugal force from the star's rotation. In Figure 7-11, the solid line is for a nonrotating dead star. The dashed line is for a dead star rotating at its breakup speed. Stable dead stars correspond to the shaded region in between these two limiting curves. As expected, when rotation is included the limiting masses of white dwarfs and neutron stars are increased, but not by very much. As a result of these kinds of calculations, astrophysicists are convinced that *no* neutron stars can exist with masses greater than about three suns.

The existence of an upper limit to the mass of a neutron star poses a dilemma. From studying binary stars, astronomers realize that stars with masses up to 40 or 50 suns exist in the sky. From calculations involving stellar evolution, astrophysicists realize that massive stars evolve very rapidly. Suppose a dying massive star does *not* eject all of its matter into space during a supernova explosion. Suppose the remaining dead core contains *more* than three solar masses. Such a star cannot become a white dwarf since its mass is far above the Chandrasekhar limit. Such a star cannot become a pulsar because its mass is too great to be supported by degenerate neutron pressure. In short, there are absolutely *no physical forces strong enough to hold up the star.* A dying star whose

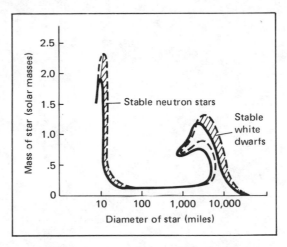

FIGURE 7-11. *Mass versus Diameter of Dead Stars.* White dwarfs are about the same size as the earth. Neutron stars are only about 10 to 15 miles in diameter.

dead core contains more than three solar masses simply gets smaller and smaller. The crushing inward weight of trillions upon trillions of tons of matter overpowers everything. As the star shrinks in size, the intensity of the gravitational field around the star gets stronger and stronger. Newtonian theory no longer gives the correct answers and astrophysicists turn to Einstein's general theory of relativity. The warping of space-time gets higher and higher as the star continues to contract. Finally, when the star has collapsed to only a few miles in diameter, space-time folds in over itself and the star disappears from the universe! What is left is called a *black hole.*

There are therefore *three* possible types of stellar corpses: white dwarfs, neutron stars, and black holes. White dwarfs have been known for many years. They are, in fact, so common that most astronomers believed that *all* dying stars become white dwarfs. But then pulsars were discovered, and now most astronomers agree that neutron stars also exist. From understanding the properties of dead stars, astronomers in the late 1960s and early 1970s began to consider the possible existence of the most bizarre type of dead star: the black hole. In this way, the discovery of pulsars stimulated a rebirth of interest in general relativity. For almost half a century after Einstein published the field equations, general relativity sat dormant. Few scientists could imagine conditions under which the effects of general relativity could become very important. In 1939, J. R. Oppenheimer and H. Snyder published a paper about the possibility of black holes, but it was largely ignored as fantasy. The prophetic work of Oppenheimer was resurrected, however, and today many astronomers believe black holes have been discovered in space.

8 THE BLACK HOLE

As early as 1795, the great French mathematician Pierre-Simon Laplace calculated that light would not be able to escape from a sufficiently massive or collapsed object. Even in Newtonian theory, if the escape velocity from an object exceeds the speed of light, the object will look *black* to an outside observer. In spite of this deduction, the idea that a *black hole* could exist in nature did not arise for almost two centuries. By the mid-1960s, however, astrophysicists had succeeded in calculating many of the details of stellar structure and stellar evolution. As a result of this new knowledge, astronomers are now aware of the fact that it is impossible to have stable dead stars with masses greater than about three suns. Since stars having much larger masses are commonly observed in the sky, astrophysicists began to consider the existence of black holes scattered throughout the universe.

As discussed in the previous chapter, a black hole is one of the three possible final products of stellar evolution. Unlike white dwarfs and neutron stars, however, a black hole is empty space. A black hole is all that is left following the catastrophic gravitational collapse of a massive dying star. As the star collapses inward, the intensity of gravity above its surface becomes so enormous that space-time folds over the star, which vanishes from the universe. All that remains is a very highly warped region of space-time.

One of the best ways of understanding the properties of black holes is to examine how objects and light rays move in these regions of extreme space-time curvature. For example, consider a black hole as shown in Figure 8-1. Imagine shining beams of light past the black hole. A beam of light passing very far from the black hole is deflected only slightly from its usual straight-line path. Very far from a black hole, space-time is almost perfectly flat, and light rays in such regions travel along straight lines. This is an important point. In spite of the contents of certain books recently published on black holes, whose accuracy is questionable, these objects pose no threat to us. Black holes cannot go around the universe randomly swallowing up planets, stars, and galaxies. Only a few thousand miles above a ten- or twenty-solar-mass black hole, space-time is quite flat, and relativistic effects are unimportant. Indeed, if one night the sun quietly (and magically!) turned into a black hole, you asleep in your home would notice nothing unusual . . . at least not until the following morning. Dawn would never come, but the earth would continue in its orbit at a distance of 93 million miles just as it has done for the past five billion years.

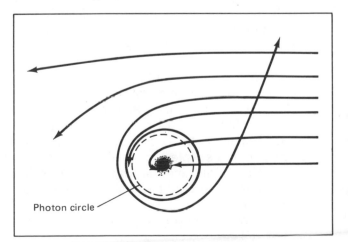

Photon circle

FIGURE 8-1. *Light Rays Passing a Black Hole.* Light rays are deflected by the intense gravitational field surrounding a black hole. Beams of light passing far from the hole are bent only slightly. Beams of light passing near the black hole can be captured into circular orbits or sucked into the hole.

Returning to Figure 8-1, notice that beams of light that pass close to the black hole are deflected through much larger angles. As light rays pass through more highly warped regions of space-time, the paths are more severely curved. In fact, it is possible to shine a beam of light toward the black hole at precisely the right distance such that the light goes into *circular orbit* about the hole. This region above the black hole is often called the *photon circle* or *photon sphere* and contains light in continuous circular orbits about the hole. Every star in the universe sends at least a few beams of light at precisely the prescribed distance from the black hole so that the light is captured into these circular orbits.

It is important to realize that the circular orbits in the photon sphere are highly *unstable*. To understand what this means, consider the nearly circular orbit of the earth about the sun. The earth's orbit is stable. If you kick the earth, nothing drastic happens. However, if a beam of light in the photon sphere deviates even a tiny amount from its precise circular orbit, the light very rapidly spirals either into the black hole or back out into space. The smallest perturbation, either inward or outward, causes light to leave the photon sphere. In this sense, all the circular orbits in the photon sphere are said to be unstable.

Finally, beams of light that are aimed almost directly at the black hole get sucked into the hole. These beams of light are lost forever from the outside universe: the black hole literally gobbles them up.

The scenario presented here is for the simplest possible type of black hole. In 1916, only a few months after Einstein published his field equations, German astronomer Karl Schwarzschild discovered a solution which was later realized to be describing the geometry of space-time about an ideal black hole. This *Schwarzschild solution* is for a spherically symmetric black hole that has *mass only*. The hypothetical dying star that created the black hole neither was rotating nor possessed an electric charge or magnetic field. The matter of the dying star fell straight inward, and the resulting black hole is therefore said to be spherically symmetric. If the black hole had been formed from a rotating star, there would be a "preferred" direction, namely the axis of rotation of the hole. No such complications are included in the Schwarzschild solution. The *Schwarzschild black hole* is the simplest conceivable type of black hole, and all discussion in this and the following chapter will be restricted to this simple picture. Later chapters will deal with electrically charged and rotating black holes.

Insight into the nature of a Schwarzschild black hole can be

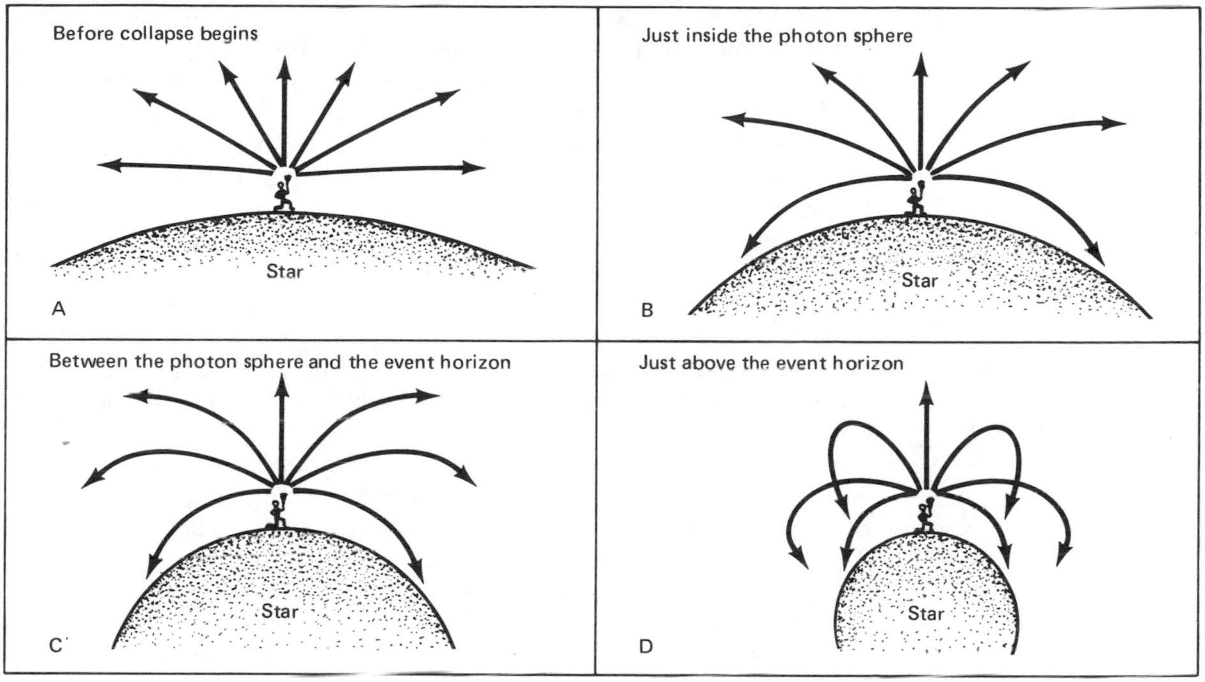
Before collapse begins
Star
A
Just inside the photon sphere
Star
B
Between the photon sphere and the event horizon
Star
C
Just above the event horizon
Star
D

FIGURE 8-2. *Beams of Light from a Collapsing Star.* An ill-fated astronaut shines beams of light from the surface of a dying star. Before the collapse begins (A), the gravitational field is comparatively weak and light rays travel along nearly straight lines. Late in the collapse (D), space-time around the star is highly warped and the light rays are severely deflected.

gained by considering a massive (uncharged, nonrotating) dying star in the process of gravitational collapse. Imagine someone standing on the surface of a dying star that has just exhausted all its nuclear fuel, as shown in Figure 8-2. Just before the collapse begins, the person takes out a powerful searchlight and shines beams in various directions. Since the matter of the star is still spread over a large volume of space, the intensity of the gravitational field above the star is still fairly weak. As a result, the light rays from the searchlight travel along straight or nearly straight lines. But once the collapse begins, the star's matter is confined into smaller and smaller volume. As the size of the star decreases, the intensity of gravity above its surface becomes stronger and stronger. The increasing curvature of space-time causes the light beams to be bent from their usual straight paths. Initially, light rays leaving the searchlight at angles near the horizon are deflected back down to the collapsing star's surface, as shown in Figure 8-2. But as the collapse proceeds, the astronaut must shine the light rays closer and closer to the vertical direction if the beams are to escape from the star. Finally, at one particular critical stage late in the process of collapse, the astronaut finds that *no* light rays manage to get away from the star. No matter what direction the astronaut points the searchlight, all the beams are deflected back down to the star. At this stage, the star is said to have fallen inside its *event horizon.* Once inside the event horizon nothing, not even light, can manage to get back out. The astronaut turns on a radio transmitter and finds that it is no longer possible to communicate with outsiders. Radio waves cannot escape from inside the event horizon, and the astronaut has literally disappeared from the universe.

The term "event horizon" is a very appropriate name for the surface in space-time from which nothing can escape. It is indeed a "horizon" beyond which no "events" can be seen. Often the event horizon surrounding a black hole is called the *surface* of the black hole.

From the Schwarzschild solution it is possible to calculate the size of the event horizon surrounding a black hole. For example, the diameter of the event horizon of a ten-solar-mass black hole is about 37 miles. As soon as a dying ten-solar-mass star collapses down to a diameter of 37 miles, space-time is so highly warped that an event horizon forms around the star. The star then disappears.

Just after a dying star falls inside its own event horizon, the star still has a substantial size, yet there are still no physical forces to hold up the star. Absolutely nothing can stop the continued contraction of the star. The entire star continues to shrink in size until finally it is crushed out of existence at a point at the center of the black hole. At this point, there is infinite pressure, infinite density, and infinite curvature of space-time. This location in space-time is called the *singularity.*

The structure of a Schwarzschild black hole may be summarized by Figure 8-3. Surrounding the entire black hole is the photon

FIGURE 8-3. *A Schwarzschild Black Hole.* A simple ideal black hole (uncharged, nonrotating) is surrounded by a photon sphere. The spherical event horizon constitutes the "surface" of the hole. The singularity is at the center of the hole.

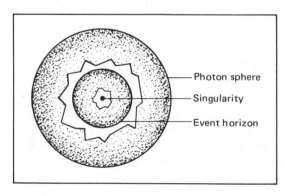

Photon sphere

Singularity

Event horizon

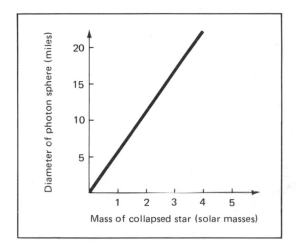

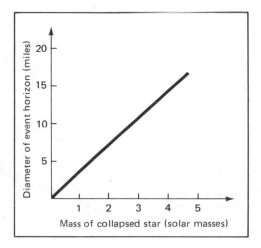

sphere consisting of light rays in unstable circular orbits. Inside the photon sphere is the event horizon, a one-way surface in space-time from which nothing can escape. Finally, the singularity is at the center of the black hole. Everything that falls through the event horizon is sucked into the singularity, where it is crushed out of existence by infinitely curved space-time. Figures 8-4 and 8-5 show the relationship between the masses of black holes and the diameters of their photon spheres and event horizons, respectively.

As a massive dying star collapses inside the photon sphere on its way to the event horizon, fewer and fewer light rays manage to escape to the outside universe. The effects depicted in Figure 8-2 become increasingly pronounced. This trapping of light rays by the collapsing star can be described with the aid of an imaginary cone called the *exit cone,* as shown in Figure 8-6. Only light rays emitted from the star at angles inside the exit cone manage to escape. Light rays leaving the star's surface at angles outside the exit cone are deflected back down to the star's surface.

As the catastrophic collapse of a massive star proceeds to its inevitable end, it becomes increasingly more difficult for rays to get away from the star's surface. Light rays have to be emitted at angles closer and closer to the vertical if they are ever to get to the outside universe. In other words, as the star falls toward the event horizon, *the exit cone closes up.* Just above the photon sphere, the exit cone is wide open. Light rays emitted at any angle can escape from the star. But by the time the star reaches the event horizon, the exit cone has become so narrow that all light rays are bent back to the star's surface.

FIGURE 8-4. *Diameter of the Photon Sphere (left).* Graph shows how the diameter of the photon sphere surrounding a Schwarzschild black hole depends on the mass of the hole. For example, a three-solar-mass black hole is surrounded by a photon sphere about 16½ miles in diameter.

FIGURE 8-5. *Diameter of the Event Horizon (right).* Diameter of the event horizon surrounding a Schwarzschild black hole depends on the mass of the hole. For example, a three-solar-mass black hole is surrounded by an event horizon about 11 miles in diameter.

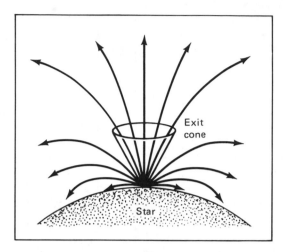

FIGURE 8-6. *The Exit Cone.* This imaginary cone aids in distinguishing between those light rays that escape from the star and those that are trapped by the star. Only those light rays emitted from the star's surface at angles inside the exit cone manage to get to the outside universe.

The behavior of the exit cone gives the first important clue to the appearance of a star forming a black hole. As the exit cone closes up, less and less light manages to leave the star. Therefore, as seen by a distant astronomer, the star appears to get dimmer and dimmer. In a realistic situation, this dimming of the dying star occurs *very* rapidly. For example, consider a ten-solar-mass star in the process of forming a black hole. As shown in Figure 8-7, the brightness decreases very rapidly as the star falls toward the event horizon. Only a hundredth of a second after the onset of gravitational collapse, the exit cone has become so narrow that less than a quadrillionth of the star's light can escape to the outside universe. It takes incredibly little time for the star to become almost totally black.

Superimposed on the rapid dimming of a dying star is a second important effect. Recall from Chapter 5 that gravity causes time to slow down. This effect is appropriately termed the *gravitational redshift* because light emitted by atoms inside a gravitational field is shifted to long wavelengths. Therefore, as the intensity of the gravitational field around the star increases during the collapse, the light from the atoms on the star's surface experiences a greater and greater redshift. Thus, as seen by a distant astronomer, a collapsing star rapidly becomes both dim and highly redshifted.

The slowing down of time, although almost undetectable in the weak gravitational field of the earth, becomes a very significant factor in the formation of a black hole. In fact, at the event horizon, time *stops* completely (see Figure 8-8). Very great care should be taken in the precise interpretation of this statement. For purposes of illustration, imagine dropping a brick down a black hole. Standing

116

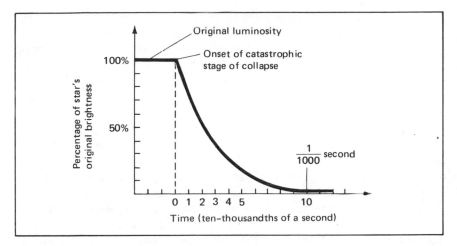

FIGURE 8-7. *The Luminosity of a Collapsing Star.* Once the final stages of collapse begin, the star becomes very dim in a very short period of time. This particular graph is for a ten-solar-mass star. After only a thousandth of a second, the star has dimmed to 2% of its original brightness. After only a hundredth of a second, brightness has declined to less than one quadrillionth its original value.

very far away from a black hole, where space-time is almost perfectly flat, you release the brick from your hands. As you watch the brick, you notice that it falls faster and faster as it plunges toward the black hole. If Newtonian theory were correct, the brick would continue to speed up and would be going almost infinitely fast when it reaches the singularity. But in such intense gravitational fields, Newtonian theory simply gives the wrong answers. Instead, as the brick approaches the event horizon, the slowing down of time begins to take over. Surprisingly, you notice that the brick starts to slow down. In fact, the brick comes to a complete stop at the event horizon because time stops at the event horizon. You, standing far

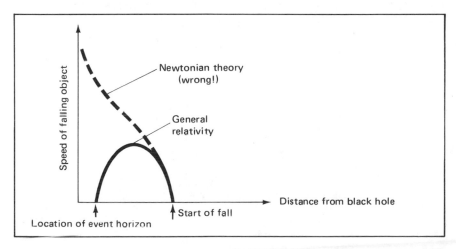

FIGURE 8-8. *The Speed of a Free-Falling Object.* As seen by a distant observer, an object falling toward a black hole appears to slow down as it nears the event horizon. The object appears to stop at the event horizon because time stops at the event horizon.

117

from the black hole, would have to wait for an infinitely long period of time to ever see the brick fall through the event horizon.

Although you never see the brick fall through the event horizon, someone falling alongside the brick tells a *very* different story. The freely falling observer does *not* notice the slowing of time. If you try to tell him that his clocks are going slow, he gives you a vigorous argument. He compares his wristwatch with other clocks in his spaceship, with the rate of decay of radioactive isotopes, and even with his own heartbeat. According to the falling observer, time is proceeding at the normal rate. The distant observer, in flat space-time, explains this peculiar situation by saying that *everything* observed by the falling observer has slowed down by precisely the same amount, even his heartbeat, his thought processes, and the rate at which he ages. According to the distant observer, the astronaut who plunges toward a black hole never reaches the event horizon. Instead, the astronaut appears destined to live forever in limbo, in a state of suspended animation, and takes billions of years to travel the last few inches up to the event horizon.

According to the falling observer, his clocks tick off time at the usual rate. He therefore falls through the event horizon in a very short time as measured by the wristwatch on his arm. Once inside the event horizon he soon realizes the error of his ways. Just as time stops at the event horizon, inside the event horizon the roles of space and time are interchanged! Far from black holes, such as here on earth, human beings have freedom to move in the three spatial directions (up-down, left-right, forward-back). But we are powerless to move around in the temporal dimension. We are dragged relentlessly forward in time, from birth, to old age, to death—whether we

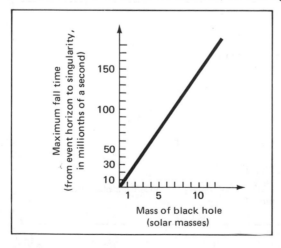

FIGURE 8-9. *The Maximum Fall Time from Event Horizon to Singularity.* A falling observer who passes through the event horizon *must* hit the singularity in a time period *less* than is given on this graph.

like it or not. Inside the event horizon, the roles of space and time interchange. Therefore, the ill-fated astronaut who plunges through the event horizon is dragged relentlessly forward *in space* toward the singularity! He is powerless to avoid striking the "crack of doom" at the singularity. Figure 8-9 shows the maximum amount of time that elapses on the astronaut's wristwatch between the event horizon and the singularity. No matter what he does, even if he possesses the most powerful rockets, he *must* strike the singularity in a period of time *less* than is given on this graph. For example, after passing through the event horizon of a 6½-solar-mass black hole, the astronaut *must* hit the singularity in *less* than a thousandth of a second.

To deal effectively with possible confusions involving the measurement of time, physicists define *two* kinds of time. *Coordinate time* is the time that an observer far from a black hole (that is, in flat space-time) measures. *Proper time* is the time that a freely falling observer measures on his wristwatch. They are *not* the same. In coordinate time, a brick falling toward a black hole takes trillions of years to reach the event horizon. In proper time, as measured by a wristwatch tied to the falling brick, the brick passes through the event horizon in a very short period. By way of an example, Figure 8-10 shows the amount of coordinate time and proper time that elapses as an object falls toward a ten-solar-mass black hole from a starting distance of 55 miles.

Concerning the matter of people falling into black holes, another interesting effect occurs which should not be overlooked. Suppose you were to fall feet-first toward a black hole. During the entire trip

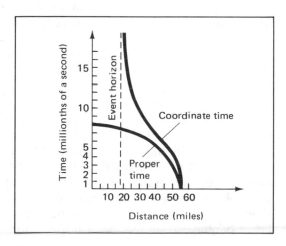

FIGURE 8-10. *Proper versus Coordinate Time.* Relationship between proper and coordinate time for an object falling into a ten-solar-mass black hole from an initial distance of 55 miles. In proper time, the infalling object strikes the singularity after only 8 millionths of a second. In coordinate time, it *never* reaches the event horizon.

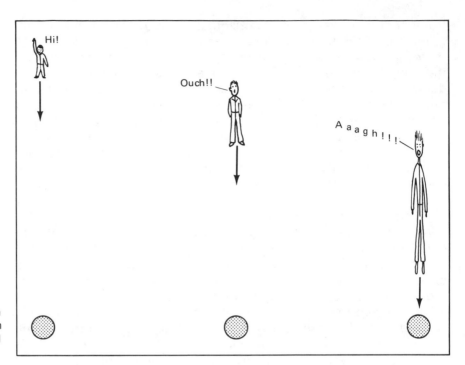

FIGURE 8-11. *Tidal Forces.* In falling toward a black hole, an observer is torn apart by tidal forces.

you are freely falling and are therefore weightless. Nevertheless, as you approach the black hole, you begin noticing something very unusual. You become aware of the fact that your feet are nearer to the black hole than your head. Your feet start falling *faster* than your head. As a result, you become stretched out in the shape of a long thin wire. By the time you reach the event horizon, you find that you are hundreds of miles tall. Obviously, falling into a black hole is not a pleasant experience. Indeed, long before you get anywhere near the photon sphere, your body will have been torn apart by incredible tidal forces. This situation is represented in Figure 8-11.

At this point, the careful reader might be a little confused. After all, if observers in flat space-time (such as astronomers here on earth) can never see anything cross the event horizon, how can black holes form in the first place? Won't it take an infinitely long time (as seen by us) for the surface of a star to reach the event horizon? Yes and no. It is indeed true that the last few atoms on the surface of a collapsing star never actually make it to the event horizon. But nobody really cares. As seen earlier in Figure 8-7, the entire star is essentially black only a few thousandths of a second after collapse begins. When the event horizon forms, virtually the entire star

120

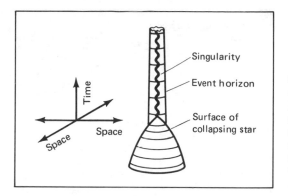

FIGURE 8-12. *The Formation of a Black Hole.* After a sufficient amount of matter has collapsed inside dimensions smaller than the Schwarzschild radius, the event horizon forms. The trapped matter then rapidly falls to the singularity at the center of the black hole.

is trapped *inside* the horizon. This matter inside the event horizon then very promptly falls into the singularity. This situation can be depicted in a three-dimensional space-time diagram as shown in Figure 8-12. In the Schwarzschild solution, the radius of the event horizon is often called the *Schwarzschild radius.* Once the required amount of matter has collapsed inside the Schwarzschild radius, the event horizon forms, thereby trapping this matter, which then collapses to a singularity. A few straggling atoms from the dying star's outer layers never make it through the event horizon and are destined to hover forever just above the Schwarzschild radius. For all practical purposes, these stragglers can be totally neglected.

One of the most fascinating and instructive techniques of understanding the structure of black holes involves taking imaginary trips in rocketships equipped with large windows. This technique will be used frequently in the next several chapters so that we can learn what adventurous astronomers would actually see as they travel up to, into, and through various types of black holes.

To introduce this technique, imagine a spacecraft, as shown in Figure 8-13. This spacecraft is equipped with two large windows. The front window is aimed directly toward the center of the black hole while the rear window faces in exactly the opposite direction, toward the outside universe. From each window exactly one-half the entire sky is seen. In addition, the spacecraft has very powerful rockets that allow the spacecraft to hover at various distances above the event horizon. Two astronomers are on board the vehicle and take photographs of what they see out of the windows at various distances from the black hole.

For convenience, the astronomers on the spacecraft choose to express their distance from the black hole in Schwarzschild radii rather than in miles or kilometers. Recall that the Schwarzschild

FIGURE 8-13. *A Spacecraft.* Two adventurous astronomers want to find out what a black hole really looks like. They therefore construct the spaceship shown here. It is equipped with two very large windows. The front window is pointed directly at the center of the black hole and the rear window faces the outside universe. One-half the entire sky can be seen from each window. The spacecraft is also equipped with powerful rockets that allow the astronomers to hover above the black hole at various distances.

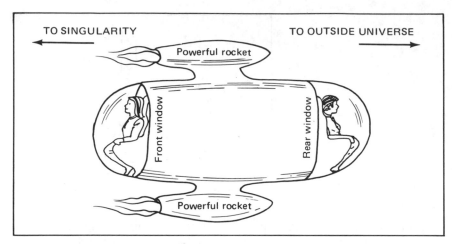

radius is defined as the radius of the event horizon. The more massive the black hole, the larger is its Schwarzschild radius. Table 8-1 gives the sizes of Schwarzschild radii for black holes of various masses. Note that the symbol Å denotes an angstrom; there are 254 million angstroms in an inch.

This table is closely related to the graph in Figure 8-5. The diameter of the event horizon of a black hole is exactly twice its Schwarzschild radius. Similarly, while the diameter of the event horizon is two Schwarzschild radii, the diameter of the photon sphere is three Schwarzschild radii.

The two astronomers in our imaginary rocketship begin their trip by simply allowing their unique spacecraft to fall straight inward toward the black hole. At various locations they turn on the powerful rockets, which momentarily stop the spacecraft. While the spacecraft is at rest, they take two photographs, one from the front window showing the view toward the black hole, and one from the rear window showing the view toward the outside universe. On five occasions the spacecraft is stopped and pairs of photographs taken:

PHOTOGRAPH	DISTANCE FROM BLACK HOLE
Photo A	Large distance (many Schwarzschild radii)
Photo B	Five Schwarzschild radii
Photo C	Two Schwarzschild radii
Photo D	At photon sphere (1½ Schwarzschild radii)
Photo E	Just above event horizon (very near one Schwarzschild radius)

TABLE 8-1   Schwarzschild Radii for Black Holes of Various Masses

Mass of Black Hole	Schwarzschild Radius (radius of event horizon)	
1 ton	13	quadrillionths Å
1 million tons	13	billionths Å
1 trillion tons	13	thousandths Å
1 quadrillion tons	13	Å
1 earth mass	⅓	inch
1 Jupiter mass	9¼	feet
1 solar mass	1.84	miles
2 solar masses	3.68	miles
3 solar masses	5.52	miles
5 solar masses	9.20	miles
10 solar masses	18.4	miles
50 solar masses	92.0	miles
100 solar masses	184	miles
1,000 solar masses	1,840	miles
One million solar masses	10	light-seconds
One billion solar masses	2.8	light-hours
One trillion solar masses	117	light-days
One quadrillion solar masses	320	light-years

Figure 8-14 shows the locations of the spacecraft above the black hole where the photographs are taken.

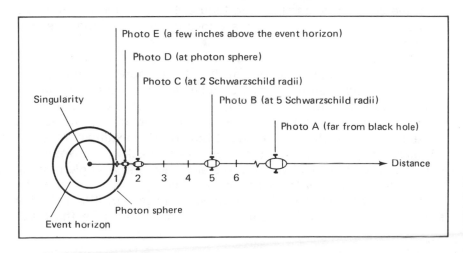

FIGURE 8-14. *A Spacecraft near a Black Hole.* Five pairs of photographs of a Schwarzschild black hole are taken at different locations.

At very great distances from the black hole, the hole looks like a very small dot seen at the center of the front window (see Figure 8-15A). The rest of the sky is essentially undistorted with one important exception. All the stars in the universe send at least a few beams of light very near the photon sphere. These beams of light circle the black hole one or more times and then spiral out toward the spacecraft. Therefore, the astronomer looking through the front window sees multiple images of all the stars in the universe piled up around the apparent "edge" of the black hole. (To avoid cluttering up Figure 8-15 A through E, the appearance of multiple images has been left out.) The view of the sky near the black hole is therefore very complicated and very distorted.

In 1975, C. T. Cunningham at Caltech performed a series of calculations that give the appearance of a black hole at various distances. Based on these computations, Figure 8-15B shows the view at five Schwarzschild radii. Since the spacecraft is now near the black hole, the hole appears larger than in Figure 8-15A. At five Schwarzschild radii (for example, at 92 miles from a ten-solar-mass black hole), the hole has an angular diameter of about 56°. The view from the rear window is still fairly undistorted.

At two Schwarzschild radii (for example, at 37 miles from a ten-solar-mass black hole), the hole looms ominously in front of the spacecraft. The angular diameter has now increased to 136°, as shown in Figure 8-15C. The rest of the sky seen from the front window is highly distorted and filled with multiple images of many stars and galaxies. Major distortions are now seen from the rear window.

At the photon sphere (for example, about 28 miles from a ten-solar-mass black hole), the black hole fills the entire forward view from the spacecraft, as shown in Figure 8-15D. Numerous multiple images are now seen around the edges of the view from the rear window.

As the spacecraft moves in toward the event horizon, the black hole appears around the edges of the rear window. The entire outside universe is seen in a small circular region centered about the outward direction from the hole, as shown in Figure 8-15E. The size of this circular region is the same as the angular size of the exit cone discussed earlier. At the event horizon (for example, at about 18 miles from the center of a ten-solar-mass black hole), where the exit cone closes up, every star in the sky is seen at a single point in the middle of the view from the rear window.

Recall that the spacecraft is equipped with powerful rockets that stop the infalling spacecraft at various distances above the black

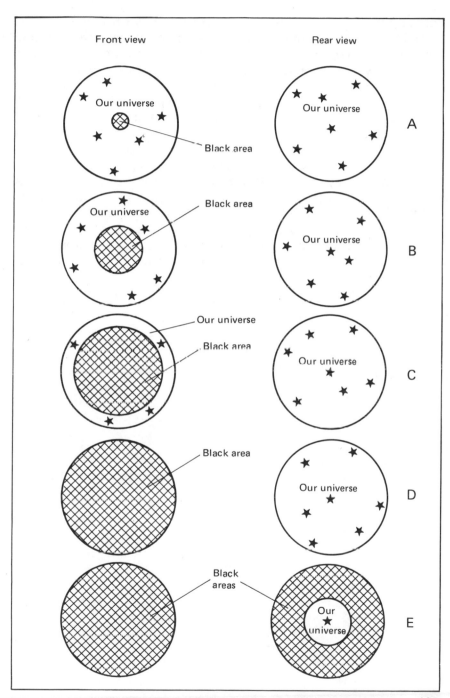

Front view   Rear view

A

Our universe — Black area

Our universe

B

Our universe — Black area

Our universe

C

Our universe — Black area

Our universe

D

Black area

Our universe

E

Black areas

Our universe

FIGURE 8-15A through E.
A: *View far from a Black Hole.*
Very far from the black
hole (a distance of many
Schwarzschild radii), the hole
looks like a small black dot in
the middle of the field of view
from the front window.

B: *View at Five Schwarzschild
Radii.* At five Schwarzschild
radii, the hole subtends an
angular diameter of about
46°, centered about the field
of view of the front window.
The distant outside universe
is still seen out of the rear
window, although some distortions are now present.

C: *View at Two Schwarzschild
Radii.* At two Schwarzschild
radii, the hole has an angular
diameter of about 136° and
fills most of the view from the
front window. The rear view
is more distorted than in B.

D: *View at the Photon Sphere.*
At the photon sphere (1½
Schwarzschild radii), the
black hole fills the entire front
view and therefore has an angular diameter of 180°. The
rear view is very distorted,
especially near the edges.

E: *View very near the Event
Horizon.* Very near one
Schwarzschild radius the entire forward view is totally
black. The apparent "edge"
of the hole has now wrapped
around into the rear view.
Through the rear window,
the entire outside universe is
now concentrated into a
small circular region centered
about the outward direction
from the hole.

hole so that the astronomers can make leisurely observations. Unfortunately the hole's gravitational field is so intense that at a few Schwarzschild radii the rockets are blasting at a furious rate. Long before the astronomers got to the point where they could take Photo B, they were subjected to accelerations of thousands of g's, which squashed them flat against the floor of the spacecraft.

To avoid being destroyed by being squashed, a second pair of astronomers conceive of *freely falling* into the black hole. Their new-and-improved spacecraft does not have any rockets to slow their fall. In addition, they and their spacecraft are miniaturized to avoid being torn apart by tidal forces. Nevertheless, they realize that this is also a suicide mission, since they are destined to strike the singularity after passing through the event horizon. This second pair of adventurous astronomers see very different views from the windows of their doomed spacecraft. However, in order to understand the meaning of what they see, we must examine the nature of the Schwarzschild geometry.

9 THE GEOMETRY OF THE SCHWARZSCHILD SOLUTION

In 1916, only a few short months after Einstein published the field equations of general relativity, the German astronomer Dr. Karl Schwarzschild discovered a mathematical solution which describes the simplest black hole. The Schwarzschild black hole is "simple" in the sense that it is spherically symmetric (that is, it has no "preferred" directions such as an axis of rotation) and contains only mass. Complications due to rotation, electric charge, or magnetic field are neglected.

Beginning in 1924, physicists and mathematicians started to realize that there was something unusual about the Schwarzschild solution to the field equations. Specifically, the mathematics of the Schwarzschild solution breaks down at the event horizon. Sir Arthur Eddington was the first to construct a new mathematical system that did not have this disadvantage. In 1933, Georges Lemaître elaborated upon these developments. However, it was

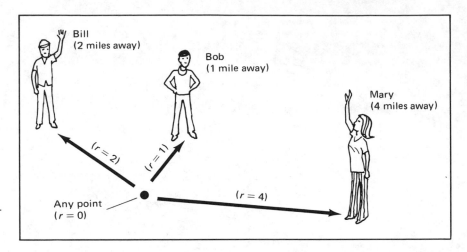

FIGURE 9-1. *Locations in Space.* The locations of objects in space can be specified by giving the radial distance from an arbitrary reference point to those objects.

John Lighton Sygne who (in 1950) discovered the true nature of the geometry of the Schwarzschild black hole, thereby paving the way for important work by M. D. Kruskal and G. Szekeres in 1960.

In order to appreciate these developments, begin by imagining three people (Bob, Bill, and Mary) floating in space, as shown in Figure 9-1. It is always possible to choose an arbitrary point in space and denote the locations of these three individuals by measuring the distances between this point and the people. For example, Bob is 1 mile away from the arbitrary reference point; Bill is 2 miles away, and Mary is 4 miles away. The letter r is usually employed in denot-

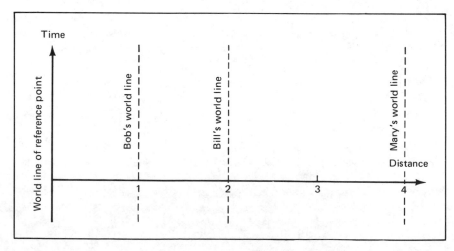

FIGURE 9-2. *A Space-Time Diagram.* A space-time diagram can be constructed in which the radial distance from an arbitrary reference point is measured along the space axis. The axes are scaled so that light rays travel along 45° lines.

ing locations in this fashion and stands for radial distance. In this way, the distance to any object in the universe can be specified.

Note, however, that the three people in this example are stationary in space but moving through time. They are getting older and older. We could therefore draw a space-time diagram, as shown in Figure 9-2. The distance from the arbitrary reference point to other points in space is measured horizontally, while time is measured vertically. Furthermore, as in special relativity, it is especially convenient to scale the axes of the graph so that light rays travel along 45° lines. In this space-time diagram, the three people in this example rise vertically. They stay at constant distances from the reference point ($r = 0$) but get older and older as time goes on.

It is important to realize that there is nothing to the left of $r = 0$ in Figure 9-2. This region corresponds to something that could be called "negative space." Since it is impossible to be "-3 feet" from the arbitrary reference point, distances from the reference point are always positive numbers.

Now consider a Schwarzschild black hole. As discussed in the previous chapter, the hole consists of a singularity surrounded by an event horizon at a distance of 1 Schwarzschild radius. A drawing (in space) of such a black hole is shown at the left of Figure 9-3. To draw a space-time diagram of this hole, the arbitrary reference point is chosen for convenience to be at the singularity. Distance is then straightforwardly measured radially outward away from the singularity. The resulting space-time diagram is shown in the right half of Figure 9-3. Just as the people (Bob, Bill, and Mary) in Figure 9-2 rise

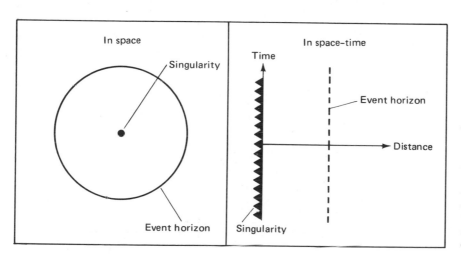

FIGURE 9-3. *A Black Hole in Space and Space-Time.* The drawing on the left shows a Schwarzschild black hole in space. The hole consists of a singularity surrounded by an event horizon. The drawing on the right shows a space-time diagram of this same hole. Distance is measured outward from the singularity.

vertically, the world line of the event horizon at exactly 1 Schwarz-schild radius from the singularity rises vertically. The world line of the singularity is shown in Figure 9-3 by a sawtooth pattern.

Although there is nothing apparently mysterious about the space-time drawing of a Schwarzschild black hole in Figure 9-3, by the 1950s physicists began to realize that such a diagram does not tell the whole story. There are two distinct regions of space-time in a black hole: those regions between the singularity and the event hori-zon and those regions beyond the event horizon. Precisely how these regions are connected is *not* properly displayed in the right half of Figure 9-3.

To see how the regions of space-time inside and outside the event horizon are connected, imagine a ten-solar-mass black hole. Imag-ine an astronomer who is ejected from the singularity, flies outward through the event horizon, goes straight up to a maximum altitude of one million miles above the hole, falls back through the event ho-rizon and (again!) hits the singularity. The astronomer and his trip are shown in Figure 9-4.

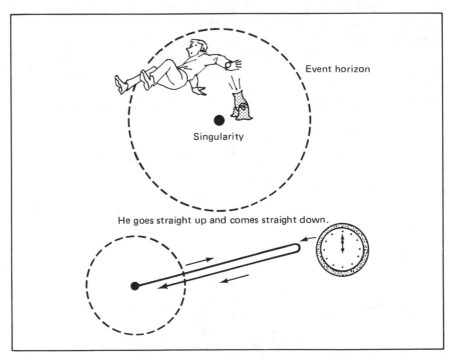

FIGURE 9-4. *An Interesting Trip.* An astronomer is ejected from the singularity of a ten-solar-mass black hole, goes through the event horizon, and rises to a maximum al-titude of one million miles. At the top of his flight, his wrist-watch (which reads proper time) is synchronized with the clocks of distant scientists (coordinate time). The as-tronomer then falls straight back into the black hole, passing through the event horizon, and hits the singu-larity.

Event horizon

Singularity

He goes straight up and comes straight down.

This may sound incredible to the careful reader, since one may ask how anybody can get out of the singularity. Suffice it to say that such a trip *is* possible mathematically. As we shall see, the full Schwarzschild solution contains *both* black holes *and* white holes. The reader's patience and indulgence are therefore required for the next few paragraphs. (The mechanism of an astronomer/astronaut traveling to a black hole will be used here and in succeeding chapters. For convenience our astronaut is usually referred to by the generic "he.")

The astronomer making this trip has a wristwatch that measures his proper time. Distant scientists observing this trip one million miles from the hole also have clocks. Their clocks in flat space-time measure coordinate time. At the top of the trajectory (one million miles from the black hole) *all* clocks are synchronized to 12 o'clock noon. It is then possible to calculate at what times (in either proper *or* coordinate time) the moving astronomer reaches various significant points along his trip.

The astronomer's wristwatch measures proper time. The wristwatch therefore does not experience the "slowing down of time" due to the gravitational redshift. Since the mass of the black hole and the maximum altitude of the trajectory are specified, calculations reveal that:

IN PROPER TIME

1. The astronomer leaves the singularity at 11:27 A.M., according to his wristwatch.
2. One ten-thousandth of a second after 11:27 A.M., he passes outward through the event horizon.
3. At 12 o'clock noon he reaches a maximum altitude of 1 million miles above the black hole.
4. One ten-thousandth of a second before 12:33 P.M., he passes inward through the event horizon.
5. The astronomer hits the singularity at 12:33 P.M.

In order words, in proper time it takes $1/10,000$ second to go from the singularity to the event horizon, either going inward or outward. It takes 33 minutes to go between the event horizon and the maximum altitude (1 million miles) either inward or outward. Notice that proper time proceeds in a normal fashion during the course of the trip.

According to the distant scientists whose clocks measure coordinate time, things are *very* different. Specifically, calculations reveal that:

IN COORDINATE TIME
1. The astronomer leaves the singularity at 11:27 A.M.
2. He passes outward through the event horizon billions of years ago (in the year − ∞).
3. At 12 o'clock noon he reaches a maximum altitude of 1 million miles above the black hole.
4. He passes inward through the event horizon billions of years in the future (in the year + ∞).
5. The astronomer hits the singularity at 12:33 P.M.

Of course, everyone agrees on the fact that the moving astronomer reaches the maximum altitude at 12 o'clock noon. That is when all the clocks are synchronized. Everyone also agrees on the times at which the astronomer leaves and hits the singularity. But in between, the pathology of the Schwarzschild geometry is apparent. In coordinate time, once the astronomer leaves the singularity, he moves *backward into the past* to the year " − ∞." He then comes forward in time, reaching the maximum altitude at noon and recrossing the event horizon in the year " + ∞." He then moves *backward in time,* arriving at the singularity at 12:33 P.M. His path in a space-time diagram is shown in Figure 9-5.

Some of these strange results can be appreciated intuitively. Recall that as seen by a distant observer (whose clocks measure coor-

FIGURE 9-5. *The Trip in Coordinate Time.* The trajectory of the moving astronomer who travels out of and into a black hole is shown in this space-time diagram. He passes outward through the event horizon in the distant past and falls back through the event horizon in the distant future.

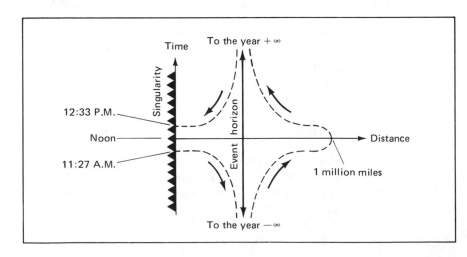

dinate time), time stops at the event horizon. Recall that as seen by a distant observer, a brick or any other object falling toward the event horizon *never* reaches the Schwarzschild radius. Therefore, the infalling astronomer cannot cross the event horizon until the year $+\infty$ in the very distant future. Since the entire trip is symmetric about 12 o'clock noon (in other words, it takes the same time to go up as to go down) the distant scientists *must* have seen the moving astronomer coming toward them for billions of years. He must have passed outward through the event horizon in the year $-\infty$.

Even more confusing is the fact that the distant scientists see *two* moving astronomers. For example, at 3 o'clock in the afternoon, they see one astronomer falling toward the event horizon (moving forward in time). But from their calculations they conclude that there *must* be another astronomer inside the event horizon falling toward the singularity (moving backward in time).

Of course, this is nonsensical. More precisely, the strange behavior of coordinate time means that the picture of a Schwarzschild black hole given in Figure 9-3 cannot possibly be correct. There must be more—perhaps a lot more—to the true space-time diagram of a black hole. In the simple picture such as Figure 9-5, somehow the same regions of space-time are covered twice; that's why two astronomers are seen when there is really only one. The goal, therefore, is to unfold or transform the simple picture in such a way as to reveal the true or *global* structure of all space-time associated with a Schwarzschild black hole.

To begin understanding what the global picture must look like, consider the event horizon. In a simple two-dimensional space-time diagram such as the right half of Figure 9-3, the event horizon is a line extending from time $-\infty$ (the distant past) to time $+\infty$ (the distant future). This line is exactly one Schwarzschild radius away from the singularity. Of course, the line really represents the location of the surface of a sphere in ordinary three-dimensional space. Surprisingly, however, when physicists try to calculate the volume of this sphere, they find that the answer is *zero.* A sphere with no volume is, of course, just a point. In other words, in the global picture of the black hole, physicists suspect that the "line" in the simple picture should really be a point!

In addition, imagine any number of astronomers jumping out of the singularity, rising to various maximum altitudes above the event horizon, and falling back in again. Regardless of precisely when they leave the singularity, and regardless of precisely how far above the black hole they get, they *all* pass through the event horizon at coordinate time $-\infty$ (on the way out) and $+\infty$ (on the way back). Therefore, clever physicists also suspect that in the global picture of the black hole, the two "points" at $+\infty$ and $-\infty$ should really be stretched into two lines!

In order to go from a simple to global picture of a black hole, the simple space-time diagram will be transformed into a much more complex space-time diagram. Nevertheless, the final result will be another space-time diagram. The spacelike quantities will be oriented in the horizontal direction (left-right) on the diagram and the timelike quantities will be oriented vertically (up-down) on the diagram. In other words, the transformation will take the *old* distance and time coordinates and convert them to *new* distance and time coordinates that will reveal the full nature of the black hole.

To gain some insight into how the old and new systems might be related, imagine someone near a black hole. In order to avoid falling into the black hole, and to remain at a constant distance from the hole, this person must have a rocketship blasting away below his feet. In flat space-time, far from any sources of gravitation, anyone in a rocketship with the motors running is experiencing an *acceleration.* He is going faster and faster as the rockets increase his speed at a steady rate. His path in a space-time diagram is shown in Figure 9-6. The path slopes closer and closer to 45° as the rockets push his speed nearer and nearer to the speed of light. The path is a curve called a *hyperbola.* Near a black hole, an observer who tries to remain at a constant distance from the hole will be experiencing a

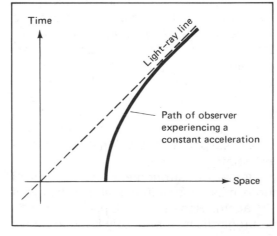

FIGURE 9-6. *An Accelerated Observer.* An accelerated object or observer (that is, one who goes faster and faster, increasing his or her speed at a constant rate) follows a hyperbolic trajectory in space-time. The trajectory gradually approaches a slope of 45° as the speed nears the speed of light.

134

continuous acceleration caused by his rocketship. Clever physicists therefore suspect that lines of "constant distance" in the new-and-improved space-time diagram will be hyperbolic.

Finally, an observer who tries to remain at rest *on* the event horizon will have to possess an incredibly powerful rocketship. In order to keep him from falling into the black hole, the rockets will have to be blasting at such a rate that the observer would be traveling with the speed of light if he were in empty space. Finally, therefore, clever physicists suspect that the event horizon lines will be sloped at exactly 45° in the new-and-improved space-time diagram.

In 1960, Kruskal and Szekeres independently discovered the appropriate transformations from the old space-time diagram of the Schwarzschild black hole to a new-and-improved space-time diagram. The new *Kruskal-Szekeres diagram* properly covers all space-time and reveals the full global structure of a black hole. All the suspicions previously discussed are confirmed, and there are a few remarkable and unexpected surprises. Although the equations of Kruskal and Szekeres convert the old picture into a new picture in one giant step, the transformation can be appreciated intuitively in the stages depicted schematically in Figure 9-7. The final result is still a space-time diagram (space-direction is horizontal; time-direction is vertical) and light rays traveling toward or away from the black hole follow 45° lines, as usual.

The final result in this transformation is very striking and almost—at first—unbelievable. Notice that there are really *two* singularities, one in the past and one in the future. In addition, there are *two* outside universes far from the black hole.

To appreciate that the Kruskal-Szekeres diagram is in fact reasonable, again consider the behavior of the moving astronomer who is ejected from the singularity, flies outward through the event horizon and later falls back in. As discussed earlier, his path in the simplified space-time diagram is very strange. This path is again shown on the left of Figure 9-8. In a Kruskal-Szekeres diagram (on the right of Figure 9-8), his path looks much more reasonable. He is really falling out of the "past" singularity and later plunges into the "future" singularity. Indeed, this so-called "analytically complete" description of the Schwarzschild solution contains *both* black and white holes. The moving astronomer actually emerges out of a white hole and later plunges into a black hole. Notice that his trip is always inclined and angles less than 45° from the vertical; his trip is always timelike and therefore allowed. Also, from comparing the left and right sides

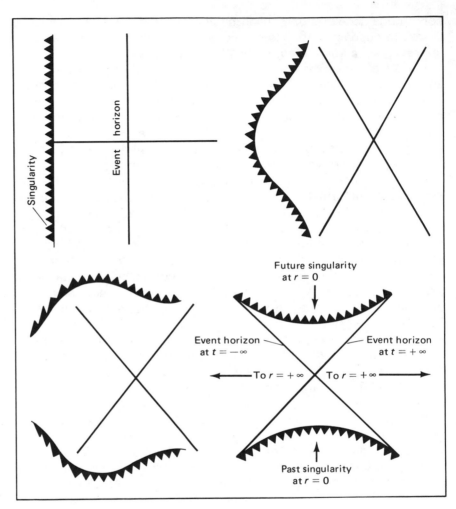

FIGURE 9-7. *The Transition to a Kruskal-Szekeres Diagram.* The transition from the old, simple-minded space-time diagram of a black hole to the new-and-improved diagram of Kruskal and Szekeres is shown schematically. The final result has *two* singularities and *two* outside universes.

of Figure 9-8, notice that the "points" at time $+\infty$ and time $-\infty$ on the event horizon have indeed been stretched into two straight lines sloped at 45°, in keeping with earlier suspicions.

In transforming to the Kruskal-Szekeres diagram, the true nature of all space-time around a Schwarzschild black hole is revealed. In the simplified picture, different regions of space-time are piled on top of each other. This is why the distant scientists watching an astronomer fall into (or out of) a black hole incorrectly thought that there were *two* astronomers. In the Kruskal-Szekeres diagram, these

136

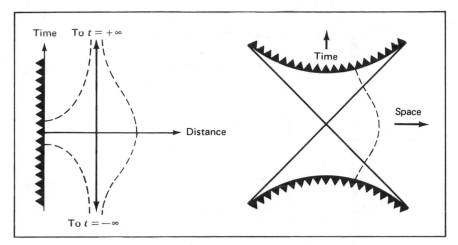

FIGURE 9-8. *Trajectories Out from and Into a Black Hole.* In a simplified space-time diagram (left), the path of an astronomer who falls out of and then into a black hole is complicated. In a Kruskal-Szekeres diagram (right), this same path is easy to understand. The astronomer is really falling out of the past singularity and into the future singularity.

overlapping regions have been properly unfolded. How these various regions in the two types of diagrams are related is shown in Figure 9-9. There are really two outside universes (Regions I and III) and two inner regions (Regions II and IV) between the singularities and the event horizon.

Finally, it is enlightening to examine how the details of a space-time grid transform from the simplified picture to the Kruskal-Szekeres diagram. In the simple case, shown in Figure 9-10, lines of constant distance (dashed) above the singularity are just straight lines rising vertically in the diagram. Lines of constant time (dotted)

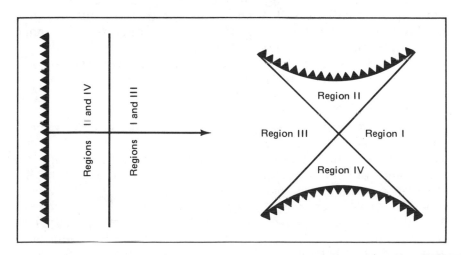

FIGURE 9-9. *Regions of Space-Time.* In the simple picture, different regions of space-time are piled on top of each other. In the Kruskal-Szekeres picture, the various regions are properly separated.

137

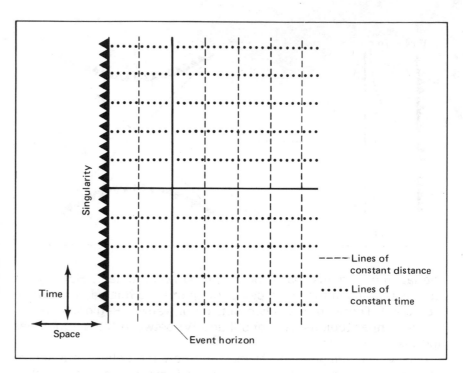

FIGURE 9-10. *Space-Time Grid for the Simple Picture.* In the simple case, lines of constant distance (dashed) above the black hole are simply straight lines that rise vertically in the diagram. Lines of constant time (dotted) are straight and horizontal.

are straight and horizontal. The space-time grid looks like a piece of common graph paper.

In a Kruskal-Szekeres diagram, shown in Figure 9-11, lines of constant time (dotted) are still straight lines but are oriented at various angles. In keeping with earlier suspicions, lines of constant distance from the black hole (dashed) are hyperbolic.

In examining Figure 9-11, it is possible to see why the roles of space and time are interchanged upon crossing the event horizon, as mentioned in the previous chapter. Notice that in the simple picture (Figure 9-10) lines of constant distance are vertical. For example, one particular dashed line might represent "ten miles above the black hole for all time." Such a line must be parallel to the event horizon in the simple picture. It must be vertical. Since it is for "all time," lines of constant distance point in the timelike direction (that is, *upward*) in the simple picture.

In Figure 9-11, showing the Kruskal-Szekeres picture, far away from the black hole the dashed lines of constant distance still point generally upward. They are still timelike. But inside the event horizon, the dashed lines of constant distance are oriented generally in

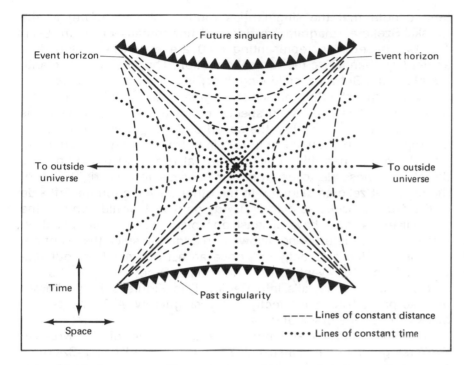

FIGURE 9-11. *Space-Time Grid in a Kruskal-Szekeres Diagram.* Lines of constant time (dotted) are straight, but lines of constant distance (dashed) are hyperbolic. The role-reversal of space and time upon crossing the event horizon is displayed.

the horizontal direction. Inside the event horizon, lines of constant distance point in the spacelike direction! Therefore, anything usually associated with distance in the outside universe behaves like time inside the event horizon.

Similarly, in the simple picture (Figure 9-10), lines of constant time are horizontal; they are pointed in the spacelike direction. For example, one particular dotted line might represent "3 o'clock in the afternoon at all locations." Such a line must be parallel to the space axis of the graph in the simple picture. It must be horizontal.

In Figure 9-11, showing the Kruskal-Szekeres picture, far away from the black hole the dotted lines of constant time still point generally in the spacelike direction (that is, horizontal). But inside the event horizon, the dotted lines of constant time point generally upward. They are oriented generally in the timelike direction. Inside the event horizon, lines of constant time point in the timelike direction! Therefore, anything usually associated with time in the outside universe behaves like distance inside the event horizon. The roles of space and time are interchanged upon crossing the event horizon.

In connection with this discussion of space and time, it is important to note that the singularities (both past and future) in the Kruskal-Szekeres diagram are oriented horizontally in Figure 9-11. The two hyperbolas representing $r = 0$ are everywhere sloped at angles *less* than 45° to the vertical direction. They are spacelike and, therefore, the Schwarzschild singularity is said to be *spacelike*.

The fact that the Schwarzschild singularity is spacelike has some profound implications. As in special relativity (see Figure 1-9) it is impossible to go faster than the speed of light and therefore spacelike trips are forbidden. It is impossible to travel along paths that are tilted by more than 45° from the vertical (timelike) direction. It is therefore impossible to go from our universe (on the right side of the Kruskal-Szekeres diagram) to the other universe (on the left side of the Kruskal-Szekeres diagram). Any path that would connect the two universes must have at least one place where it is spacelike. Such a path is therefore not allowed. In addition, since the event horizon is oriented at exactly 45°, once an astronomer from our universe falls into the event horizon, he can never get out again. For example, once anyone falls into Region II in Figure 9-9, *all* allowed timelike paths lead him directly to the singularity. A Schwarzschild black hole is "the crack of doom."

In order to appreciate more fully the nature of the Kruskal-Szekeres geometry, it is instructive to slice up the Kruskal-Szekeres diagram along spacelike sheets. These sheets will provide *embedding diagrams* of the warping of space around a black hole. This technique of slicing space-time along spacelike hypersurfaces was employed earlier in Figures 5-9, 5-10, and 5-11, and gave important insight into the shape of space near the sun.

Figure 9-12 shows a Kruskal-Szekeres diagram sliced along five typical spacelike hypersurfaces. Slice A is at an early time. Initially, the two universes outside the black hole are disconnected. The spacelike slice strikes the singularity in going from one universe to the other. Therefore, the embedding diagram for Slice A has two disconnected universes (represented by two parallel asymptotically flat sheets) each of which contains a singularity. At a later time, as both universes evolve, the singularities join and produce a bridge which does not contain a singularity. This corresponds to Slice B, which does not touch the singularity. As time moves on, the bridge or *wormhole* opens up and reaches a maximum diameter of two Schwarzschild radii at the time of Slice C. Still later, the bridge gets narrower (Slice D) and finally pinches off (Slice E), leaving the universes again disconnected. This evolution of a wormhole (Figure 9-12) occurs in less than $1/10,000$ second for a solar-mass black hole.

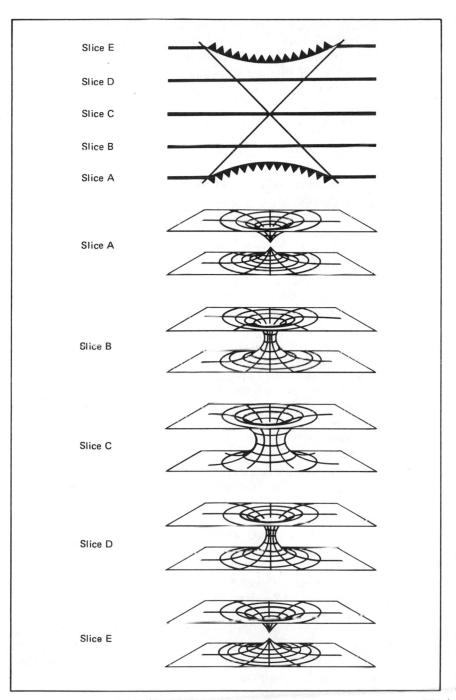

Slice E

Slice D

Slice C

Slice B

Slice A

Slice A

Slice B

Slice C

Slice D

Slice E

FIGURE 9-12. *Embedding Diagrams of a Black Hole.* To construct embedding diagrams, the Kruskal-Szekeres geometry is sliced along five representative spacelike hypersurfaces. The evolution of the resulting wormhole is revealed in going from Slice A (at an early time) to Slice E (at a later time).

The discovery of the global structure of space-time of a black hole by Kruskal and Szekeres constituted a fundamental breakthrough for theoretical astrophysicists. For the first time, diagrams could be drawn that completely display all regions of space and time. Since 1960, however, some important improvements have been made, primarily by Roger Penrose. In a Kruskal-Szekeres diagram, although the whole story is presented, the diagram extends forever to both the right and left. For example, our universe goes on forever to the right of the Kruskal-Szekeres diagram while the "other" parallel, asymptotically flat universe goes on forever to the left of the diagram. Penrose realized that it would be very instructive and useful to "map" these large regions into smaller areas where we see exactly what is going on far from a black hole. To accomplish this feat, Penrose invented a powerful technique of *conformal mapping,* whereby all of space-time, including entire universes, can be drawn in a single diagram.

To introduce Penrose's techniques, consider ordinary flat space-time, such as was shown in Figure 9-2. All of space-time is on the right side of the diagram simply because you can never be at negative distances from an arbitrary reference point. You can be 6 feet away from a reference point, but never -6 feet. Think carefully about Figure 9-2. The world lines of Bob, Bill, and Mary are shown only over a limited region of space-time because of the size of the page. If you want to see where Bob, Bill, and Mary go to in one thousand years or where they were a billion years ago, you would need a much longer piece of paper. It would be very useful if those locations far from the here-and-now regions could be mapped into a neat, small diagram.

The most distant regions of space-time are appropriately called the *infinities.* Such regions are extremely far away in space or time (that is, in the distant future or distant past). As shown in Figure 9-13, there are five kinds of infinities. First of all, there is I^-, the *past timelike infinity.* This is where all material objects (Bob, Bill, Mary, the earth, galaxies, etc.) came from. All material objects travel along timelike paths and must all go to I^+, the *future timelike infinity,* billions of years from now. I^0 is *spacelike infinity,* and since nothing can travel faster than the speed of light, nothing (except perhaps tachyons) can ever get to I^0. While nothing known to scientists can travel faster than light, photons travel at exactly the speed of light along 45° lines in a space-time diagram. It is therefore possible to define $\mathscr{I}^-$, *past null infinity,* which is where all light rays came from. Finally, $\mathscr{I}^+$, *future null infinity,* is where all light rays go to. All distant

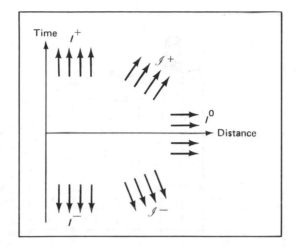

FIGURE 9-13. *The Infinities.*
The farthest reaches of
space-time can be divided up
into five kinds of infinities.
Past timelike infinity (I^-) is
where all material objects
come from and future
timelike infinity (I^+) is where
they go to. Past null infinity
($\mathscr{I}^-$) is where light rays come
from and future null infinity
($\mathscr{I}^+$) is where they go to.
Nothing (except tachyons)
can get to spacelike infinity (I^0).

regions of space-time belong to one of the five infinities: I^-, $\mathscr{I}^-$, I^0, $\mathscr{I}^+$, or I^+.

Penrose's method consists of a mathematical technique of squeezing all the infinities onto a single piece of paper. The transformation equations that accomplish this feat act like bulldozers (shown symbolically in Figure 9-14) which pile up the most distant reaches of space-time and bring them in where they can be studied. The final result is shown in Figure 9-15. Note that lines of constant distance from the arbitrary reference point are essentially vertical and point generally in the timelike direction. Lines of constant time are essentially horizontal and point generally in the spacelike direction.

In the *conformal map* of all flat space-time (Figure 9-15), everything is squeezed inside a triangle. All of past timelike infinity (I^-) is concentrated into a single point at the bottom of the diagram. All timelike paths of all material objects originate at this point representing the very distant past. All of future timelike infinity (I^+) is concentrated into a single point at the top of the diagram. The timelike paths of all material objects in the universe ultimately end up at this point representing the distant future. Spacelike infinity (I^0) is concentrated at a point to the right of the diagram. Nothing (except tachyons) ever gets to I^0. The past and future null infinities ($\mathscr{I}^-$ and $\mathscr{I}^+$) are transformed into 45° lines that bound the lower right and upper right of the diagram. All light rays always travel along 45° lines, and light coming from the distant past originates along $\mathscr{I}^-$ while light going into the distant future ultimately approaches $\mathscr{I}^+$.

143

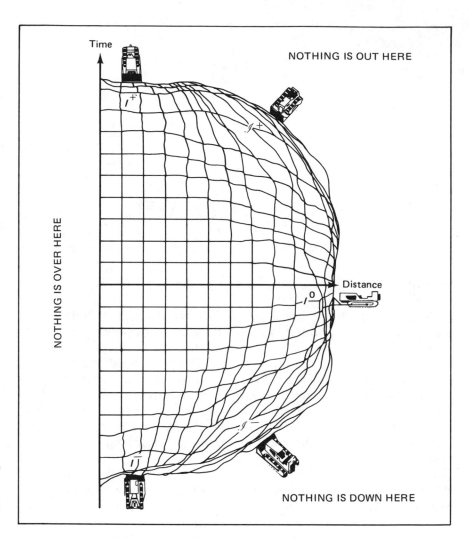

FIGURE 9-14. *Penrose Conformal Mapping.* Mathematical techniques can be used to squeeze the most distant reaches of space-time (the five infinities) into a small and convenient area.

The vertical straight line that bounds the left of the diagram is simply the timelike world line of the arbitrary reference point ($r = 0$).

To finish the introduction to a Penrose conformal diagram of flat space-time, Figure 9-16 shows the complete world lines of Bob, Bill, and Mary. Compare this diagram with Figure 9-2. They are really the same. In the conformal diagram, however, the entire extent of world lines (from the distant past at I^- to the distant future at I^+) is seen.

Ordinary flat space-time is not terribly exciting or interesting. However, Penrose's techniques can be applied to black holes. Spe-

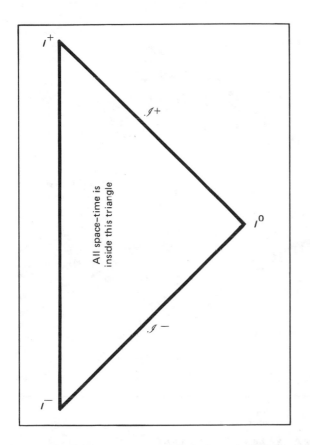

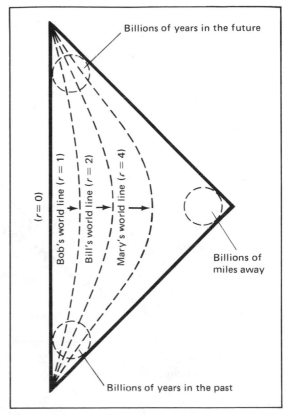

I^+

$\mathscr{I}^+$

All space-time is
inside this triangle

I^0

$\mathscr{I}^-$

I^-

Billions of years in the future

$(r = 0)$

Bob's world line $(r = 1)$

Bill's world line $(r = 2)$

Mary's world line $(r = 4)$

Billions of
miles away

Billions of years in the past

cifically, the Kruskal-Szekeres diagram (see Figure 9-11) can be mapped conformally so that the physicist can see *all* of space-time in all the universes displayed on a single sheet of paper. As shown symbolically in Figure 9-17, Penrose's mathematical transformation equations again act like little bulldozers that push in all of space-time. The final result is shown in Figure 9-18.

In the Penrose diagram of a Schwarzschild black hole (Figure 9-18) again notice that the behavior of lines of constant time and lines of constant distance is essentially the same as in the Kruskal-Szekeres diagram. The event horizon is still oriented at 45° and the singularities (past and future) are still spacelike. The role-reversal of space and time still occurs upon crossing the event horizon. Now, however, the most distant parts of both universes associated with the black hole are seen. The five infinities of our universe (I^-, $\mathscr{I}^-$, I^0, $\mathscr{I}^+$, I^+) appear on the right of the diagram and the five infinities of

FIGURE 9-15. *A Penrose Diagram of Flat Space-Time* (*left*). All space-time is squeezed into a triangle by Penrose's technique of conformal mapping. Three of the infinities (I^-, I^0, I^+) are concentrated into points. The null infinities ($\mathscr{I}^-$, $\mathscr{I}^+$) become straight lines oriented at 45°.

FIGURE 9-16. *An Example of a Penrose Conformal Diagram* (*right*). This diagram is really the same as Figure 9-2. But a conformal diagram shows the entire world lines of objects (from the distant past at I^- to the distant future at I^+).

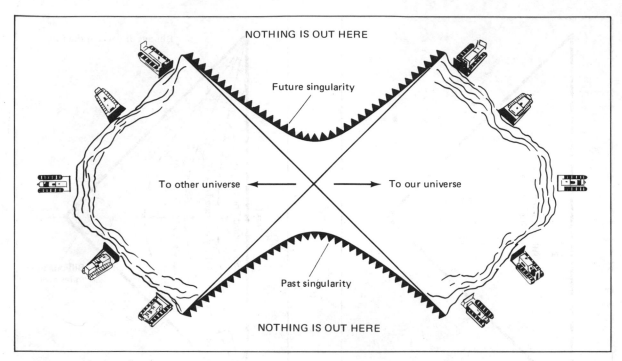

NOTHING IS OUT HERE

Future singularity

To other universe ← → To our universe

Past singularity

NOTHING IS OUT HERE

FIGURE 9-17. *Conformal Mapping of a Black Hole.* All of space-time associated with a Schwarzschild black hole can be conformally mapped onto one page using Penrose's techniques. These techniques push all of space-time into regions where it can be seen and studied.

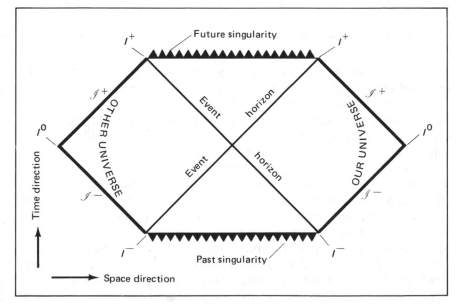

Future singularity

I^+ I^+

$\mathscr{I}^+$ OTHER UNIVERSE   Event   horizon   OUR UNIVERSE $\mathscr{I}^+$

I^0 Event   horizon I^0

Time direction

$\mathscr{I}^-$ $\mathscr{I}^-$

I^- I^-

Past singularity

Space direction

FIGURE 9-18. *The Penrose Diagram of a Schwarzschild Black Hole.* This diagram is essentially the same as the Kruskal-Szekeres diagram shown in Figure 9-11. Now, however, the most distant regions of the two universes (I^-, $\mathscr{I}^-$, I^0, $\mathscr{I}^+$, I^+ for *each* universe) are seen.

146

the other universe (I^-, $\mathscr{I}^-$, I^0, $\mathscr{I}^+$, I^+) are lined up on the left of the diagram.

As a final exercise with Schwarzschild black holes, we are now in a position to understand what some adventurous (and suicidal) astronomers see as they plunge into a black hole and pass *through* the event horizon.

The spaceship used by the astronomers is shown in Figure 9-19. The front window always faces exactly toward the singularity while the rear window faces in the opposite direction, back toward us in the outside universe. Note that the spacecraft is *not* equipped with rockets to slow its inward free fall. Starting from a great distance above the black hole, the astronomers simply fall straight in at an ever-increasing speed (as measured by them). Their path, shown in Figure 9-20, takes them first through the event horizon and then into the singularity. Since their speed is always less than the speed of light, their trajectory on a Penrose diagram must be timelike: it must always slope at angles less than 45° from the vertical.

During the trip, the astronomers take four pairs of photographs, one from each of the two windows at various locations. The first pair of photographs (A) is taken when they are still very far from the black hole. As seen in Figure 9-21A, the distant black hole looks like a small dot at the center of the front view. Although the sky is distorted very near the black hole, the rest of the sky appears quite normal. As the astronomers fall faster and faster toward the hole, the distant universe seen out of the rear window becomes increasingly redshifted.

Although distant astronomers claim that the infalling spacecraft slows down and stops at the event horizon, the astronomers *in* the spacecraft notice no such effect. According to the moving as-

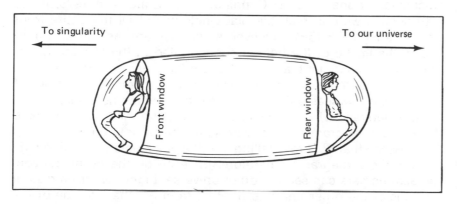

FIGURE 9-19. *A Spaceship.* Two adventurous (and ill-fated) astronomers use this spacecraft for a trip into a black hole. Note that the spaceship has no rockets to slow its inward free fall. The front window faces the center of the hole while the rear window faces the outside universe.

147

FIGURE 9-20. *The Suicide Mission.* The fatal path followed by the astronomers is shown here on a Penrose diagram. During the course of the trip, they take four pairs of photographs. The first photographs (A) are taken far from the black hole. The second (B) are taken at the instant they cross the event horizon. The third (C) are taken in between the event horizon and the singularity. The final photographs (D) are taken just before striking the singularity.

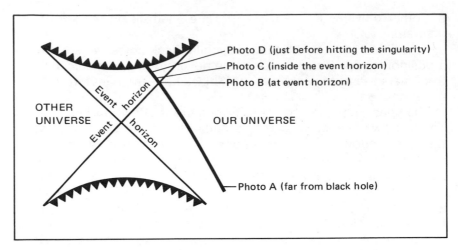

tronomers, their speed is constantly increasing. In fact, as measured by the falling astronomers, as they pass through the event horizon their speed is a sizable fraction of the speed of light. This is important because the moving astronomers therefore observe an aberration of starlight very similar to that discussed in Chapter 3 (see Figures 3-9, 3-10, 3-11). Recall that someone moving at near the speed of light sees major distortions of the visible sky. Specifically, the images of objects in the sky tend to appear in front of the moving observer. As a result of this effect, the image of the black hole is concentrated toward the center of the front window of the infalling spacecraft.

The views seen by the moving astronomers at the event horizon are shown in Figure 9-21B. This and succeeding views are based on calculations done by C. T. Cunningham at Caltech in 1975. If the astronomers were at rest, the black hole would fill the entire front view (see Figure 8-15E). However, since they are moving at a high speed, the image of the hole is concentrated in the forward direction. It has an angular diameter of about 80°. Images of the sky near the black hole are very distorted and the astronomer looking out of the rear window sees only the universe from which he came.

To understand what is seen *inside* the event horizon, refer back to the Penrose diagram of a Schwarzschild black hole (Figure 9-18 or 9-20). Recall that light rays falling into the black hole travel along 45° lines in these diagrams. Therefore, once inside the event horizon, the astronomers can see the other universe. Light rays from distant portions of the other universe (that is, from $\mathscr{I}^-$, the left side of the

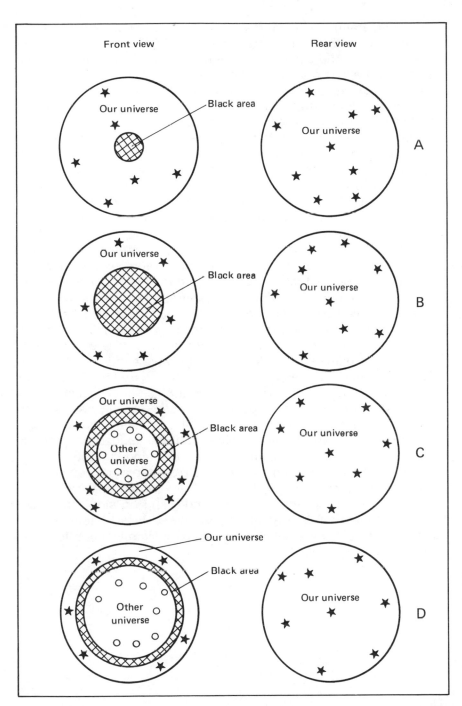

Front view   Rear view

Our universe   Black area

A

Our universe

Our universe   Black area

B

Our universe

Our universe   Black area
Other universe

C

Our universe

Our universe   Black area
Other universe

D

Our universe

FIGURE 9-21A through D.
A: *Far from the Black Hole.* At a great distance, the black hole looks like a small black dot in the middle of the field of view of the front window. From the rear window, the infalling astronomers see an undistorted view of where they came from.

B: *At the Event Horizon.* Due to aberration effects, the image of the black hole is concentrated toward the center of the front window. The astronomer looking out of the rear window sees only the universe from which he came.

C: *Between the Event Horizon and the Singularity.* Once inside the event horizon, the astronomer looking out of the front window can see the other universe. Light from $\mathscr{I}^-$ in the other universe occupies the central regions of her field of view.

D: *Just above the Singularity.* As the astronomers approach the singularity, more and more of the other universe is seen from the front window. The image of the black hole itself (a ring) becomes thinner and thinner and rapidly approaches the edge of the front view.

Penrose diagram) can reach the astronomers. As shown in Figure 9-21C, in between the event horizon and the singularity, the other universe is seen at the center of the view from the front window. The black region of the hole itself now appears as a *ring* separating the view of our universe from the view of the other universe.

As the infalling astronomers approach the singularity, the black ring becomes thinner and thinner and approaches the edge of the field of view from the front window. The appearance of the sky just above the singularity is shown in Figure 9-21D. More and more of the other universe is seen from the front window. In fact, at the singularity, the other universe completely fills the front view. During this entire trip, the astronomer looking out of the rear window sees only our outside universe, although the images have become increasingly distorted.

There is one important effect noticed by the infalling astronomers which is not displayed on the "photographs" represented in Figures 9-21A, B, C, D. Recall that light escaping to the distant universe from near the event horizon is highly redshifted. This phenomenon is called the *gravitational redshift* and was discussed in Chapters 5 and 8. When light coming from an intense gravitational field is redshifted, it loses energy. Conversely, light falling *into* a black hole is blueshifted; it gains energy. A weak radio wave coming in from the distant universe, for example, is converted into an intense x-ray or gamma ray just above the event horizon. If black holes described by Penrose diagrams such as Figure 9-18 *really* exist in nature, then light falling inward from $\mathscr{I}^-$ has been piling up near the event horizon for billions of years. This infalling light has gained tremendous amounts of energy and therefore when the infalling astronomers pass through the event horizon they encounter a sudden and intense blast of x-rays and gamma rays. The light that has been coming in from $\mathscr{I}^-$ of the other universe and piling up along the event horizon is called the *blue sheet.* As will be seen in Chapter 13, the existence of such blue sheets has profound implications for grey holes and white holes.

10 ELECTRICALLY CHARGED BLACK HOLES

From understanding the details of stellar evolution, astronomers realize that black holes might exist in our galaxy and throughout the universe. In the previous two chapters, we have discussed numerous properties of the simplest black hole as described by the Schwarzschild solution to the field equations. The Schwarzschild black hole contains only mass. It has no electric charge. It does not possess a magnetic field. It is not rotating. All the properties of a Schwarzschild black hole are uniquely determined *only* by the mass of the dying star whose gravitational collapse created the hole.

The Schwarzschild solution is obviously an oversimplification. At the very least, a *real* black hole would probably be rotating. But exactly how complex is a real black hole? What complexities should be included or neglected in a complete description of a black hole that we might expect to observe in the sky?

Imagine a massive star that has just used up all its nuclear fuel and is just about to begin a catastrophic gravitational collapse. This star would conceivably be very complicated. A detailed description of such a star would include a wide range of complexities. In principle, the astrophysicist could calculate the chemical composition throughout the star, or how the temperature varies from the star's center to its surface, or the precise state of matter (for example, the pressure and density) at various depths inside the star. These computations would be difficult and would critically depend on the complete history of the star. Stars formed from different clouds of gas at different times would be expected to have different internal structures.

In spite of all these complications, there is one thing of which we can be sure. As long as the mass of the dying star is greater than about three solar masses, it *must* become a black hole at the end of its life cycle. No physical forces exist that can prevent the ultimate final collapse of a massive star.

To see what this means, recall that a black hole is a highly warped region of space-time out of which absolutely nothing—not even light—can escape. In other words, *no information can get out of a black hole.* Once the event horizon forms about a dying massive star, it becomes impossible to discover any intricate details about what might be occurring inside the event horizon. Information that might be contained inside the event horizon is forever lost from the universe. A black hole is therefore sometimes called an *information sink.*

Although great quantities of information are lost when a star collapses down to become a black hole, some information still remains behind. For example, the highly warped space-time surrounding the hole indicates that a star has died. Specific properties of the hole such as the diameter of the photon sphere or event horizon, as shown in Figures 8-4 and 8-5, are directly related to the mass of the dead star. Even though the hole is literally black, a space traveler far from the hole is readily aware of the existence of its gravitational field. By measuring how much the trajectory of the spacecraft is deflected from a straight line, the traveler can calculate precisely the total mass of the black hole. The mass of a black hole is therefore one piece of information that is *not* lost.

To illustrate this point further, imagine two identical stars collapsing to form black holes. On one star we place a ton of bricks. On the other star we place a one-ton elephant. After the black holes have formed, we can measure the intensity of the gravitational fields at large distances just by observing the orbits of planets or satellites. We would find that the fields have identical strengths. At very great distances from the holes, Newtonian mechanics and Kepler's laws can be used to calculate the total masses of the two holes. Since the total masses of the ingredients of the holes were the same, the final results are the same. More importantly, we cannot tell which hole swallowed the elephant and which swallowed the bricks. That sort of detailed information is lost forever. If you drop a ton of *anything* down a black hole, the final result is always the same. It is possible to discover how much matter went in, but its shape, color, chemical composition, and so on, are gone.

It is always possible to measure the total *mass* of a black hole, because the gravitational field of the hole influences the geometry of space and time for vast distances around the hole. A physicist far from the black hole can perform experiments to measure this gravitational field, for example, by launching satellites and observing their orbits. This important insight allowed physicists to present strong arguments about what is *not* swallowed by black holes. Specifically, anything that can be measured by a hypothetical scientist far from the hole has *not* been completely swallowed.

When James Clerk Maxwell began working on electromagnetic theory in the mid-nineteenth century, he had great quantities of information concerning electric and magnetic fields at his disposal. One surprising fact was that electric and magnetic forces depend on distance in exactly the same way that gravity depends on distance. Both gravitational and electromagnetic forces are *long range;* their influences can be felt very far from their sources. By contrast, the forces that hold the nuclei of atoms together, strong and weak interactions, are *short-range* forces. These nuclear forces are felt over a very tiny region surrounding nuclear particles.

The long-range nature of electromagnetic forces means that a physicist far from a black hole can perform experiments to tell whether or not the hole is *charged.* If the black hole has an electric charge (either positive or negative) or a magnetic charge (either north pole or south pole), the distant physicist can use sensitive apparatus to detect these charges. Therefore, in addition to mass, the *charge* on a black hole is not lost information.

There is a third and final important effect that can be measured by the distant physicist. As we shall see in the next chapter, any rotating object tries to drag space-time around with it. This phenomenon is known as the *Lense-Thirring effect* or the "dragging of inertial frames." To a very small extent our earth drags space and time around as it rotates. This phenomenon is more pronounced for rapidly rotating massive objects. If a black hole is formed from a rotating star, the dragging of space and time around the hole should be quite noticeable. Physicists sitting in a rocketship far above the hole will notice that they are gradually pulled around the hole in the direction the hole is rotating. The closer they are to the rotating black hole, the faster they are pulled around.

In discussing any rotating object, physicists often speak of the *angular momentum* of the object. Angular momentum is simply a quantity related to the mass of the object and its rate of rotation. The faster it rotates, the more angular momentum it has. Therefore, in addition to mass and charge, the angular momentum of a black hole is not lost information.

During the late 1960s and early 1970s, theoretical astrophysicists worked very hard on the question of what properties are swallowed or retained by a black hole. The result of their efforts is the famous "No Hair" theorem first proposed by John Wheeler at Princeton. We have seen that the mass of a black hole, its charge, and its angular momentum are quantities that can be measured by a distant observer. These three basic quantities are retained by the black hole and characterize the geometry of space-time about the hole. From the work of Stephen Hawking, Werner Israel, Brandon Carter, David Robinson, and others, it has been shown that these are the *only* properties not swallowed by a black hole. In other words, if the mass, charge, and angular momentum of a black hole are given, you know everything there is to know: Black holes have no properties except mass, charge, and angular momentum. Black holes are therefore very simple objects. For example, they are much simpler than the stars from which they were created. A complete description of a star involves all sorts of things such as chemical composition, pressures, densities, and temperatures. No such complications are involved in black holes (Figure 10.1). Indeed, *black holes have no hair!*

Since black holes are completely described by three parameters (mass, charge, and angular momentum), there must be a small number of solutions to the Einstein field equations characterizing every conceivable type of black hole. For example, in the previous two chapters, we examined the simplest kind of black hole: a hole that has only mass whose geometry is given by the Schwarzschild solution. This solution was discovered in 1916, and although numerous other solutions for mass-only black holes have been discovered since, they are *all* equivalent to the Schwarzschild solution.

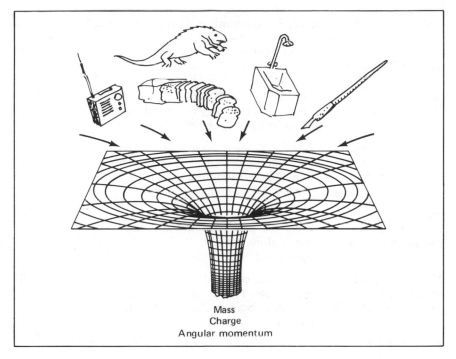

FIGURE 10-1. *"Black Holes Have No Hair!"* Virtually all information about objects falling into a black hole is lost forever. Only the mass, charge, and angular momentum of the infalling objects are not "eaten" by the hole. This means that black holes are very simple objects. They can be completely described by only three parameters: mass, charge, and angular momentum. (Adapted from John Wheeler)

It is impossible to conceive how black holes could form without matter. Therefore every black hole must have some mass. But, in addition to mass, the hole could have charge or rotation or both. Between 1916 and 1918, H. Reissner and G. Nordstrøm discovered a solution to the field equations for a black hole possessing mass and charge. The next major advance was delayed until 1963, when Roy P. Kerr discovered a solution for a black hole having mass and angular momentum. Finally, in 1965, Newman, Couch, Chinnapared, Exton, Prakash, and Torrence published their solution of the most complex type of black hole, namely one that has mass, charge, and angular momentum. Each of these solutions is unique; there are no other possibilities. At most, a black hole has only three *parameters:* mass (symbolized by M), charge (either electric or magnetic, symbolized by Q) and angular momentum (symbolized by a). All the possible solutions are summarized in Table 10-1.

With the addition of each new complexity (either charge or rotation, or both) the geometry of space-time is severely affected. The Reissner-Nordstrøm and the Kerr solutions are vastly different from each other and from the Schwarzschild solution. Of course, in the

TABLE 10-1   Solutions to Field Equations Characterizing Black Holes

Type of Black Hole	Description of Black Hole	Name of Solution	Year Discovered
Mass only (parameter M)	The "simplest" black hole. Has mass only. Spherically symmetric.	Schwarzschild solution	1916
Mass and charge (parameters M, Q)	A charged black hole. Has mass and charge (electric or magnetic). Spherically symmetric.	Reissner-Nordstrøm solution	1916 and 1918
Mass and angular momentum (parameters M, a)	A rotating black hole. Has mass and angular momentum. Axially symmetric.	Kerr solution	1963
Mass, charge and angular momentum (parameters M, Q, a)	A rotating, charged black hole. The "most complex" black hole. Axially symmetric.	Kerr-Newman solution	1965

limit of vanishing charge and angular momentum ($Q \rightarrow 0$ and $a \rightarrow 0$) all three complex solutions must reduce down to the Schwarzschild solution. Nevertheless, black holes that have charge and/or angular momentum possess some amazing properties.

During World War II, H. Reissner and G. Nordstrøm discovered a solution to the Einstein field equations which completely describes a "charged" black hole. Such a black hole can have an electric charge (either positive or negative) and/or a magnetic charge (either north pole or south pole). While electrically charged objects are commonplace, magnetically charged objects are not. Things that have magnetic fields (for example, a magnet, a compass needle, the earth) have *both* north and south poles. In fact, until very recently most physicists believed that magnetic poles always occur in pairs. However, in 1975 a team of scientists from Berkeley and Houston announced that they had discovered a *magnetic monopole* in one of their experiments. If their results are confirmed, individual magnetic charges can exist. A magnetic north pole could exist without a south pole, or vice versa. The Reissner-Nordstrøm solution allows for the possibility that a black hole could have a monopole magnetic field.

Regardless of the details of how a black hole got its charge, all of the properties of the charge are lumped together in one single number, Q, in the Reissner-Nordstrøm solution. This is analogous to the fact that the Schwarzschild solution does not depend on how a black hole got all its mass. The mass could have come from elephants, bricks, or stars; the final result is always the same. Furthermore, the geometry of space-time in the Reissner-Nordstrøm solution does not depend on the nature of the charge. The charge could be positive, negative, north pole, or south pole. The only thing that matters is the total amount of charge, which is written as $|Q|$. Thus, a Reissner-Nordstrøm black hole depends on only two numbers: the total mass of the hole M and the total charge on the hole $|Q|$.

In thinking about real black holes that might exist in the universe, physicists realize that the Reissner-Nordstrøm solution is probably *not* very important, because electromagnetic forces are much stronger than gravitational forces. For example, the electric field of an electron or proton is trillions upon trillions of times stronger than its gravitational field. This means that *if* a black hole did have a huge charge, the enormous strength of its electromagnetic field would rapidly tear apart gases and atoms floating in space. In a very short period of time particles of the same charge as the hole would be strongly repelled while particles of opposite charge would be vigorously attracted by the hole. By violently attracting particles of opposite charge, the hole would soon be neutralized. Therefore, in reality, black holes would be expected to have only weak charges. The value of $|Q|$ must be very much smaller than M for a real black hole. Indeed, calculations reveal that for a black hole that might exist in space, M must be at least a hundred billion billion times bigger than $|Q|$. Stated mathematically: $M >> |Q|$.

In spite of the (nasty!) conditions imposed on us by physical reality, it is very instructive to examine the Reissner-Nordstrøm solution in detail. This will pave the way for an in-depth discussion of the Kerr solution in the next chapter.

To introduce some of the features of the Reissner-Nordstrøm solution, imagine an ordinary uncharged black hole. According to the Schwarzschild solution, such a black hole consists of a singularity surrounded by an event horizon. The singularity is at the center of the hole (at $r = 0$) and the event horizon is at 1 Schwarzschild radius (at $r = 2M$). Now imagine adding a very small amount of charge to this black hole. As soon as charge is introduced, we must turn to the Reissner-Nordstrøm solution for the geometry of space-time. According to the Reissner-Nordstrøm solution, there are *two* event ho-

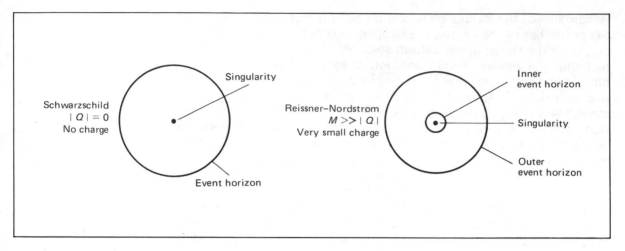

FIGURE 10-2. *Charged and Uncharged Black Holes.* The addition of even a tiny amount of charge results in the appearance of a second (inner) event horizon just above the singularity.

rizons. There are two locations above the singularity where—as seen by a distant observer—time stops. With the addition of a tiny amount of charge, the event horizon that was at 1 Schwarzschild radius moves inward slightly. More surprisingly, a second event horizon appears just above the singularity. There is both an *outer event horizon* and an *inner event horizon* surrounding the singularity of a charged black hole. The structure of an uncharged (Schwarzschild) black hole and that of a charged (Reissner-Nordstrøm with $M >> |Q|$) black hole are shown in Figure 10-2.

If more and more charge is added to a black hole, the outer event horizon shrinks while the inner event horizon increases in size. Finally, when so much charge has been piled on the black hole that $M = |Q|$, the two horizons merge. If still more charge is added, the event horizons disappear completely and we are left with a *naked singularity.* For $M < |Q|$ there are *no* event horizons and the singularity is exposed to the outside universe. This condition violates the famous "law of cosmic censorship" proposed by Roger Penrose. This idea states that "Thou shalt not have naked singularities!" and will be discussed more fully later in the text. The sequence of diagrams in Figure 10-3 shows the locations of the event horizons for black holes with the same mass but different charges.

While Figure 10-3 shows the locations of event horizons and the singularity of black holes *in space,* it is instructive to examine space-time diagrams of charged black holes. To construct such diagrams, which are graphs of time versus distance, we adopt the straight-forward method that was introduced in the previous chapter (see

158

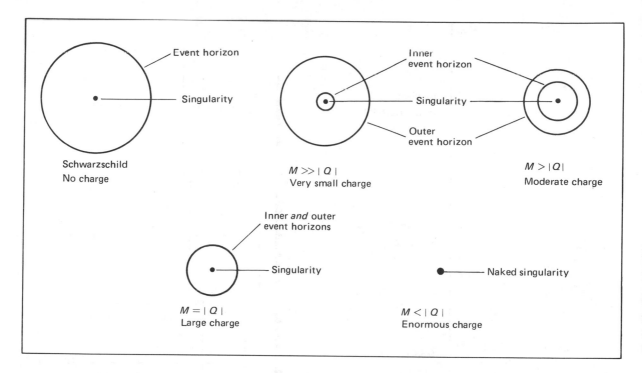

Event horizon

Singularity

Schwarzschild
No charge

Inner
event horizon

Singularity

Outer
event horizon

$M \gg |Q|$
Very small charge

Singularity

$M > |Q|$
Moderate charge

Inner *and* outer
event horizons

Singularity

$M = |Q|$
Large charge

Naked singularity

$M < |Q|$
Enormous charge

Figure 9-3). Distance is measured outward from the singularity and is plotted horizontally. As usual, time is plotted vertically. In such a diagram, the singularity always borders the left side of the graph and runs vertically from the distant past to the distant future. The world lines of the event horizons are also vertical and separate the outside universe from the innermost regions of the black hole.

Figure 10-4 shows space-time diagrams for a series of black holes all having the same mass but different charges. For comparison, the diagram of a Schwarzschild black hole is shown at the top of the sequence. (Recall that the Schwarzschild solution is exactly the same as the Reissner-Nordstrøm solution with $|Q| = 0$.) Notice that when a very little charge is added to a black hole, a second (inner) event horizon appears close to the singularity. For a black hole with moderate charge ($M > |Q|$), the inner event horizon is located further from the singularity, while the outer has been pulled further inward. In the case of a very large charge ($M = |Q|$, sometimes called the *extreme Reissner-Nordstrøm solution*), the two event horizons merge. And finally, for an enormous charge ($M < |Q|$), all

FIGURE 10-3. *Charged Black Holes in Space.* As more and more charge is added to a black hole, the outer event horizon shrinks while the inner event horizon grows. When so much charge has been added that $M = |Q|$, the two horizons merge. For still higher charges, the event horizons vanish, leaving an exposed or naked singularity.

159

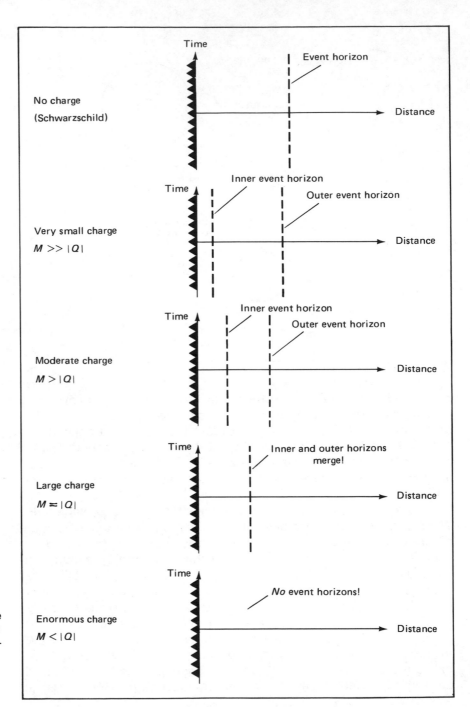

FIGURE 10-4. *Space-Time Diagrams of Charged Black Holes.* This sequence of diagrams shows the appearance of space-time for black holes having the same mass but different charges. For comparison, the diagram of a Schwarzschild hole ($|Q| = 0$) is shown at the top.

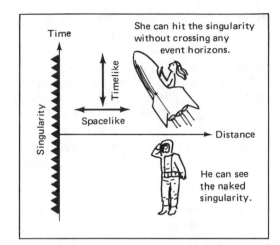

FIGURE 10-5. *A Naked Singularity.* No event horizons surround a black hole that has an enormous charge $(M < |Q|)$. In violation of the "law of cosmic censorship," the singularity is exposed to the outside universe for all to see.

event horizons vanish. As depicted in Figure 10-5, with no event horizons the singularity is exposed to the outside universe. Distant observers can see the singularity and astronauts can fly right up to this region of infinitely curved space-time without ever crossing through an event horizon. Detailed calculations reveal that gravity is *repulsive* very near the singularity. Although a distant astronaut is attracted by the black hole, very near the singularity the astronaut is repelled. In sharp contrast to the Schwarzschild solution, space very near a Reissner-Nordstrøm singularity is a region of antigravity.

Two event horizons and repulsive gravitational forces near the singularity are not the only surprises contained in the Reissner-Nordstrøm solution. From the detailed discussion of the Schwarzschild solution, recall that diagrams like Figure 10-4 do *not* tell the whole story. Recall that in the Schwarzschild geometry we encountered severe difficulties due to the fact that simplified graphs actually had *several* regions of space-time piled on top of each other (see Figure 9-9). The same kinds of difficulties exist in the diagrams in Figure 10-4, and we must now proceed to straighten things out.

The complete picture of the *global structure* of space-time can be obtained by following a few very simple rules. We have already seen what the global structure of a Schwarzschild black hole looks like. This was shown in Figure 9-18 and is called a *Penrose diagram.* This drawing may be called the Penrose diagram of a Reissner-Nordstrøm black hole for the special case of *no* charge ($|Q| = 0$). Furthermore, we realize that if charge is taken away from a Reissner-Nordstrøm hole (as $|Q| \rightarrow 0$), whatever diagrams we come up with

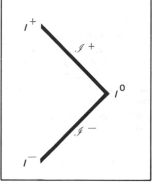

Far away from anything, the conformal structure of space–time must have a triangular appearance with the five infinities . . . like this:

FIGURE 10-6. *An Outside Universe.* In any Penrose diagram of any kind of black hole, the outside universes must each have a triangular shape with the five infinities (I^-, $\mathscr{I}^-$, I^0, $\mathscr{I}^+$, I^+). The resulting outside universe can be pointed either to the right (as shown here) or to the left.

must collapse down to the Penrose diagram for the Schwarzschild case. The first rule therefore is: There must be another universe opposite our universe which is accessible only along prohibited space-like trips.

In constructing a Penrose diagram of a charged black hole, it is perhaps reasonable to expect a number of universes. All such universes must each have the five kinds of infinities (I^-, $\mathscr{I}^-$, I^0, $\mathscr{I}^+$, I^+) discussed in the previous chapter. Furthermore, each of these outside universes must have a triangular appearance since Penrose's technique of conformal mapping acts like little bulldozers (recall Figure 9-14 or Figure 9-17) which push all of space-time into a small triangular shape. The second important rule therefore is: Any and all outside universes must have a triangular appearance with the five infinities. This outside universe can be pointed either to the right (as shown in Figure 10-6), or to the left.

To appreciate the third rule, recall that the event horizon of a Schwarzschild black hole is oriented at 45° in a Penrose diagram (see Figure 9-18). The third rule becomes: Any and all event horizons must be lightlike and therefore are always oriented at 45°.

To obtain the fourth and final rule, recall that the roles of space and time are interchanged in crossing the event horizon of a Schwarzschild black hole. Detailed analyses of spacelike and timelike directions reveal that the same effect occurs with charged black holes. The fourth rule therefore states: The roles of space and time interchange *every time* an event horizon is crossed.

Figure 10-7 illustrates this fourth rule for the case of small or moderate charge ($M > |Q|$). Far from the charged black hole, the spacelike direction is parallel to the space axis and the timelike direction is parallel to the time axis. In crossing through the outer event horizon, these two directions are interchanged; the spacelike direction is now parallel to the time axis and the timelike direction is parallel to the space axis. But, moving still further inward through the inner event horizon, there is another role-reversal. The orientation of spacelike and timelike directions near the singularity as well as far from the charged black hole is the same.

This *double* role-reversal of spacelike and timelike directions has an important consequence concerning the nature of the charged singularity. In the uncharged Schwarzschild black hole there is only *one* role-reversal of space and time. Inside the lone event horizon, lines of constant distance are pointed in the spacelike (horizontal) direction. Thus the line representing the location of the singularity

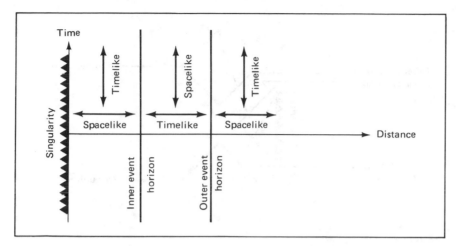

FIGURE 10-7. *Role-reversals of Space and Time (M > |Q|).* Every time an event horizon is crossed, the roles of space and time are interchanged. This means that in a charged black hole, there are *two* complete role-reversals of space and time because there are two event horizons.

($r = 0$) must be horizontal; it must be pointed in the spacelike direction. However, when there are *two* event horizons, lines of constant distance near the singularity point in the timelike (vertical) direction. Therefore the line representing the location of a charged singularity ($r = 0$) must be vertical; it must be pointed in the timelike direction. We therefore arrive at the very important conclusion that the singularity of a charged black hole must be timelike!

Using the above cookbook rules, it is now possible to construct a Penrose diagram of the Reissner-Nordstrøm solution. Begin by imagining an astronaut sitting on the earth in our universe. He blasts off in his rocketship and heads toward a charged black hole. As shown in Figure 10-8, our universe in a Penrose diagram has a triangular appearance with the five infinities. The astronaut's (allowable) timelike trip must always be pointed at angles less than 45° from the vertical because he can never travel faster than light. Allowed paths are shown as dashed lines in Figure 10-8. As the astronaut approaches the charged hole, he passes through the outer event horizon, which must be oriented at exactly 45°. Once inside, the astronaut can never return to *our* universe. But he can fall inward through the inner event horizon, which is also oriented at 45°. Once inside the inner event horizon, the astronaut can (foolishly) crash into the singularity where he experiences repulsive gravitational forces and infinite warping of space-time. Notice, however, that a tragic end to the spaceflight is *not* imperative! Because the singularity of a charged black hole is timelike, it must be vertical in the

163

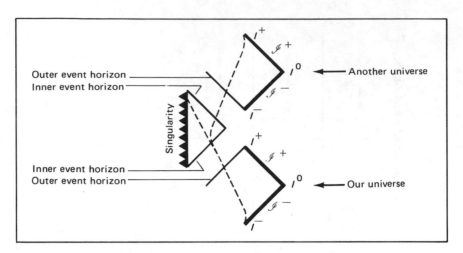

FIGURE 10-8. *A Piece of a Penrose Diagram.* By considering possible trips of an astronaut traveling from our universe into a charged black hole, part of the Penrose diagram for the Reissner-Nordstrøm solution can be constructed.

Penrose diagram. The astronaut could avoid death simply by piloting his vehicle away from the singularity along an allowed timelike path, as shown in Figure 10-8. This safe trip takes him away from the singularity and he passes back out through the inner event horizon, which again must be at 45°. Continuing on, the astronaut passes through the outer event horizon (again oriented at 45°) and emerges into an outside universe. Since this trip obviously takes time, the sequence of events along the trip must be spread out into the future. The astronaut therefore does *not* re-emerge into *our* universe, but rather comes out into a separate *future* universe. As expected, this future universe must have a triangular appearance in the Penrose diagram with the usual five infinities.

It should be emphasized that in constructing this Penrose diagram, we are again encountering both black and white holes. The astronaut can travel outward through event horizons and emerge in a future universe. Most physicists strongly believe that white holes cannot exist in nature. Nevertheless, we shall continue with this theoretical treatment of the global structure of space-time which involves black and white holes back-to-back. Arguments against the existence of white holes will be deferred until Chapter 14.

This little story and the drawing in Figure 10-8 must be only part of the whole picture. There must be more to a Penrose diagram of a charged black hole because, at the very least, there has to be yet another universe opposite to ours which is accessible only along (prohibited) spacelike trips. This conclusion is based on the realiza-

164

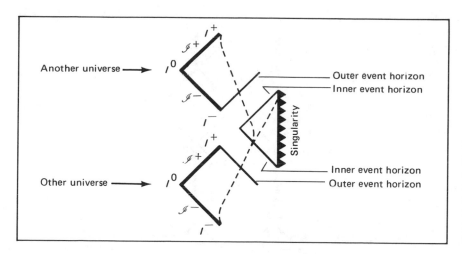

Another universe ⟶

Other universe ⟶

Outer event horizon
Inner event horizon
Singularity
Inner event horizon
Outer event horizon

FIGURE 10-9. *Another Piece of a Penrose Diagram.* By considering possible trips of an alien astronaut from the other universe, another part of the Penrose diagram for the Reissner-Nordstrøm solution can be constructed.

tion (Rule 1) that if all the charge is taken away, the Penrose diagram must reduce to the Schwarzschild solution. Although no one in our universe could ever get to this "other" universe because it is impossible to travel faster than light, we could imagine an astronaut from the other universe traveling toward the same charged black hole. The possible trips are shown in Figure 10-9.

The trip of the alien astronaut from the other universe sounds just like the trip of the astronaut who blasted off from the earth in our universe. The other universe has the usual familiar triangular appearance in the Penrose diagram. In traveling toward the charged black hole, the alien astronaut passes through the outer event horizon, which must be oriented at 45°. Still later, he passes through the inner event horizon, also oriented at 45°. The alien now has a choice: either hit the timelike singularity (vertical in the Penrose diagram), or turn around and pass back through the inner event horizon. To avoid an unpleasant fate, the alien chooses to leave the black hole and emerges through the inner event horizon, which is, as usual, tilted at 45°. Still later, he moves outward through the outer event horizon (tilted at 45° in the Penrose diagram) and emerges into a new, future universe.

Each of these two hypothetical trips gives only two pieces of the entire Penrose diagram. The entire picture is obtained simply by putting the pieces together, as shown in Figure 10-10. This diagram must be repeated over and over infinitely far into the future and infinitely far into the past because either of the two astronauts dis-

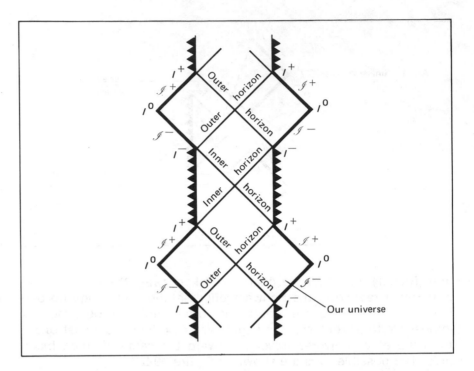

FIGURE 10-10. *The Entire Penrose Diagram for a Reissner-Nordstrøm Black Hole (M > |Q|).* The complete Penrose diagram for a black hole with a small or moderate charge (M > |Q|) is obtained by putting together the pieces from Figures 10-8 and 10-9. The Penrose diagram is repeated over and over into both the future and the past.

cussed could choose to leave the future universes into which they emerged and plunge back into the charged black hole. In doing so, they could emerge into new universes still farther into the future. Similarly we could imagine other astronauts from universes in the distant past who emerge into our universe. The complete Penrose diagram therefore is repeated over and over like a long string of paper dolls. The full, global geometry of a charged black hole therefore connects an infinite number of past and future universes with our own! Equally surprising is the fact that astronauts could use a charged black hole to make spaceflights into these other universes. This incredible conclusion is closely related to the concept of a *white hole,* which will be discussed in a later chapter.

The treatment of the global structure of space-time just discussed applies to the case of black holes with small or moderate charge (M > | Q |). However, in the extreme Reissner-Nordstrøm black hole (M = | Q |), there is so much charge that the inner and outer event horizons merge. This merging of the two event horizons has some interesting consequences.

Recall that far from a charged black hole (outside the outer event

166

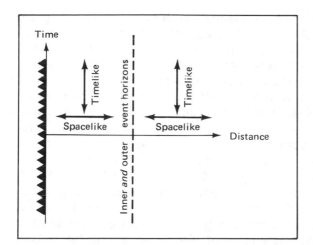

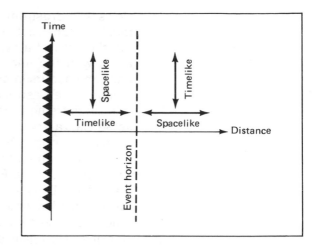

horizon) the spacelike direction is parallel to the space axis and the timelike direction is parallel to the time axis. Also recall that near the singularity (inside the inner event horizon; after *two* complete role-reversals of space and time), the spacelike direction is again parallel to the space axis and the timelike direction is again parallel to the time axis. As more and more charge is added to a Reissner-Nordstrøm black hole, the region *between* the two event horizons gets smaller and smaller. When, finally, so much charge has been added that $M = |Q|$, this in-between region is squashed down to nothing. Thus, in passing through the merged inner-and-outer event horizon, there is *no* role-reversal of space and time. Alternatively, the reader can think of two role-reversals occurring simultaneously at the single event horizon in the extreme Reissner-Nordstrøm black hole. Either way, the final result is the same. As shown in Figure 10-11, the timelike direction is everywhere parallel to the time axis and the spacelike direction is everywhere parallel to the space axis.

Although there is only one event horizon for the extreme Reissner-Nordstrøm black hole, the situation is *very* different from a Schwarzschild black hole, which also has only one event horizon. With a single event horizon, there is always a role-reversal of spacelike and timelike directions, as shown in Figure 10-12. But in the extreme Reissner-Nordstrøm black hole, the event horizon can be thought of as being "double." The inner and outer event horizons are on top of each other. The final result is that there is no net role-reversal.

That the inner and outer event horizons merge in an extreme

FIGURE 10-11. *A Space-Time Diagram for an Extreme Reissner-Nordstrøm Black Hole* ($M = |Q|$) (*left*). When so much charge has been added that $M = |Q|$, the inner and outer event horizons merge. This means that there is no role-reversal of space and time in crossing the resulting (double) event horizon.

FIGURE 10-12. *The Schwarzschild Black Hole* ($|Q| = 0$) (*right*). Although the (uncharged) Schwarzschild black hole has only one event horizon, there is a role-reversal of space and time from one side to the other. Compare this with Figure 10-11.

Reissner-Nordstrøm black hole means that a *new* Penrose diagram must be constructed. As before, this can be done simply by following the path of a hypothetical astronaut. The same cookbook rules apply with the important exception that *no* role-reversal of space and time occurs in crossing the event horizon.

Imagine an astronaut who blasts off from the earth and plunges into an extreme Reissner-Nordstrøm black hole. As usual, our universe has a triangular appearance in the Penrose diagram. After passing through the event horizon, the astronaut is faced with a choice. He can hit the singularity, which must be timelike and therefore vertical in the Penrose diagram. Or, as shown in Figure 10-13, he can pilot his spacecraft away from the singularity along an allowed timelike path. If the latter course is chosen, the astronaut emerges at some later time through the event horizon into another universe. He again has the choice of remaining in this future universe and visiting some planets, or turning around and going back into the hole. If the astronaut goes back in, he can thread his way upward in the Penrose diagram to any number of future universes.

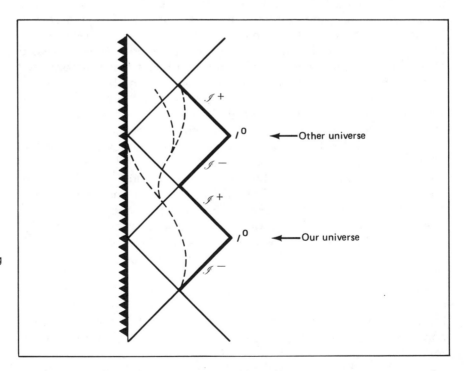

FIGURE 10-13. *The Penrose Diagram for an Extreme Reissner-Nordstrøm Black Hole* ($M = |Q|$). By examining the possible paths of an astronaut plunging in and out of an extreme Reissner-Nordstrøm black hole, a diagram of the global structure of space-time can be constructed.

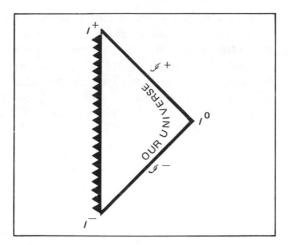

FIGURE 10-14. *A Naked Singularity*. A naked singularity ($M < |Q|$) has no event horizons. A black hole of this type does *not* connect our universe to any other universes.

The complete picture is shown in Figure 10-13. As before, the diagram is repeated over and over infinitely far into the past and the future like a long chain of paper dolls.

Although physically unreasonable, it is mathematically possible for a black hole to have so much charge that $M < |Q|$. In this case, the charge is so enormous that all event horizons disappear and we are left with a naked singularity. Since there are no event horizons, there are no questions of any role-reversals of space and time. The singularity just sits there for all to see. A *naked singularity* is an exposed region of infinitely warped space-time, unclothed by any event horizons.

An astronaut who leaves the earth and plunges toward a naked singularity never passes through any event horizons. He is always in *our* universe. Near the singularity, he experiences strong repulsive gravitational forces. With powerful rockets, the astronaut can (foolishly!) hit the singularity. And that's all. A naked singularity does not connect our universe with any other universes. As is the case with all charged black holes, the singularity is timelike and therefore must be vertical in a Penrose diagram. Since there are no other universes besides our own, the Penrose diagram of a naked singularity is very simple. As shown in Figure 10-14, our universe has the usual triangular appearance with the five infinities. The singularity borders the left side of the diagram. Anything and everything to the left of the singularity is cut off from us. Nothing can get through the singularity.

169

Since real black holes can have only very tiny charges (if any at all), much of the preceding discussion is purely academic. Nevertheless, straightforward techniques for drawing complicated Penrose diagrams have been established. Although charged black holes are probably not significant in astrophysics, rotating black holes are very important. Indeed, astrophysicists suspect that real black holes could be rotating at enormous rates and therefore many of the techniques developed here will be very useful in discussing Kerr black holes.

11 ROTATING BLACK HOLES

The whole idea of black holes in space really got started when astronomers began understanding how stars evolve. In particular, in
the 1960s it was shown that a dying star whose mass is greater
than three solar masses could not be supported by any known
physical forces. Presumably such a star would collapse down to
zero volume, thereby producing a singularity in space-time surrounded by at least one event horizon. By 1970, astrophysicists
had proven that, besides mass, black holes can have no more
than two additional properties. They can have charge or angular
momentum or both. Black holes possessing only mass are described by the Schwarzschild solution and were discussed in
Chapters 8 and 9. Black holes with mass and charge (either electric or magnetic) are described by the Reissner-Nordstrøm solution
and were treated in the previous chapter. However, in considering
the consequences of charged black holes, astrophysicists realize

that strong arguments can be presented which demonstrate that real black holes cannot be highly charged. If a black hole did form with a large charge, it would be quickly neutralized by tearing apart nearby gases floating in space. Real black holes have either a very tiny charge or no charge at all.

Does this mean that real black holes, as might be found in space, are Schwarzschild holes? No! Astronomers believe that virtually all stars rotate. The sun rotates once around its axis every four weeks. Additionally, astronomers have discovered that, in general, more massive stars rotate more rapidly. These massive stars are, of course, good candidates for becoming black holes. Now recall from the discussion of pulsars (Chapter 7) that as a dying star shrinks in size, its rate of rotation increases. This is a direct consequence of the law of *conservation of angular momentum.* A collapsing star rotates more rapidly for exactly the same reason that an ice skater doing a pirouette speeds up as she pulls in her arms (see Figure 7-6). Since dying stars rotate more and more rapidly as they collapse to smaller and smaller sizes, it is entirely reasonable to suppose that a real black hole must be rotating. It must have angular momentum.

The idea that realistic models of black holes must include rotation is not new. Nevertheless, for half a century after the formulation of general relativity, the Schwarzschild solution was used in all calculations. Everyone realized that the effects of rotation should be included, but no one knew how to solve the Einstein field equations properly. Specifically, a complete solution to the field equations which includes rotation must depend on *two* parameters: the mass of the black hole (symbolized by the letter M) and the angular momentum of the black hole (symbolized by the letter a). Furthermore, such a solution must be *asymptotically flat;* far from the black hole, space-time must be flat. The field equations are mathematically so complicated that no one could find a single solution that met these simple requirements.

A major breakthrough in relativity theory occurred in 1963 when Roy P. Kerr, an Australian mathematician then at the University of Texas, discovered a complete solution to the field equations for a rotating black hole. For the first time in almost half a century since Einstein's original work, astrophysicists finally had the mathematical tools to describe the geometry of space-time surrounding a massive, rotating object. By 1975, it was proved that the *Kerr solution* is unique. Just as all possible solutions for black holes with mass (M only) are equivalent to the Schwarzschild solution, and just as all possible solutions for black holes with mass and charge (M and Q) are equivalent to the Reissner-Nordstrøm solution, all possible solutions with mass and angular momentum (M and a) must be equivalent to the Kerr solution. The discovery of the Kerr solution stands as one of the most important developments in theoretical astrophysics of the mid-twentieth century.

Prior to the work of Kerr, the only important effect known about rotating masses in general relativity was the dragging of inertial frames, sometimes called the Lense-Thirring effect, briefly mentioned in the previous chapter. The dragging of inertial frames is a phenomenon whereby space-time is pulled around by a rotating object. General arguments can be presented which show that this phenomenon must occur above any rotating object. But until the discovery of the Kerr solution in 1963, astrophysicists had no way of calculating how significant this effect would be in the case of rotating black holes. By the late 1960s, however, detailed examination of the dragging of inertial frames by black holes had led to some remarkable discoveries.

Perhaps the best way to introduce the dragging of inertial frames by black holes involves a simple experiment with flashbulbs. A flashbulb, as used in photography, emits a sudden burst of light. In ordinary flat space, this sudden burst of light expands outward equally in all directions away from the flashbulb at a speed of 186,000 miles per second. At any moment after the flash, there is an outward-expanding spherical shell of light centered precisely on the location of the flashbulb. As shown in Figure 11-1, it is possible to draw this expanding shell of light schematically as a circle centered on the location of the flash.

To appreciate the properties of black holes, imagine setting off numerous flashbulbs at various distances from a black hole. Consider, first, a static (Schwarzschild) black hole as shown in Figure 11-2. Set off flashbulbs at various distances from the black hole and examine the locations of the resulting expanding shell of light. Far from the black hole, where space-time is essentially flat, the expanding shell is always centered precisely on the location of the flashbulb at the instant the flash was emitted. But as flashbulbs are set off closer and closer to the black hole, the expanding shell is pulled more and more towards the hole. In fact, when a flashbulb is set off at the event horizon, the expanding shell of light lies entirely to the inward side of the location of the flashbulb. This must be so, because nothing, not even light, can escape from the horizon. Inside

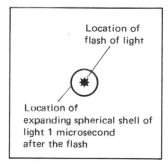

FIGURE 11-1. *A Flashbulb in Flat Space-Time.* The asterisk indicates the location of the flashbulb at the instant of the flash. The circle indicates the location of the outward-expanding spherical shell of light one microsecond after the flash. In flat space-time, the shell of light is centered on the location of the flashbulb.

173

the event horizon, light is pulled so strongly toward the singularity that the location of the flashbulb lies outside the expanding shell, as shown in Figure 11-2.

This experiment tells us that inside the event horizon of a Schwarzschild black hole it is impossible to remain at rest. Since it is impossible to travel faster than light, everything inside the event horizon is pulled toward the singularity. Furthermore, in order for you to remain at rest on the event horizon your outward speed must equal the speed of light. Again imagine an astronaut in a rocketship. As he approaches a black hole his rockets must blast furiously to prevent him from falling in. The closer he is, the more powerful his rockets must be in order for him to remain at rest above the hole. Indeed, at the event horizon, the rocket must be blasting so violently that his outward speed equals the speed of light. Otherwise he would be sucked in. Inside the event horizon, he is doomed to fall inward no matter how powerful his rockets are. The event horizon of a Schwarzschild black hole is therefore the smallest distance above the black hole where the astronaut can be at rest. In a Schwarzschild black hole, the event horizon is the *static limit*. At the static limit, one must travel at the speed of light to stay at the same place.

Now imagine repeating the flashbulb experiment with a rotating black hole. Far from the hole, where space-time is essentially flat, the expanding shell of light is centered exactly about the position of the flashbulb at the instant it was set off. But as flashbulbs are set off closer to the black hole, *two* simultaneous effects are noticed. As before, the gravitational field of the hole pulls the light inward. But since the hole is rotating, space-time is dragged around the hole. Therefore the expanding shell of light is also pulled around the hole in the direction of rotation. As shown in Figure 11-3, the combination of these effects causes the expanding shell of light to be

FIGURE 11-2. *Flashes of Light near a Schwarzschild Black Hole.* Expanding spherical shells of light from flashbulbs set off near a nonrotating black hole are pulled toward the hole. Indeed, the spherical shells of light from flashbulbs set off at or inside the event horizon expand entirely on the inward side of the location of the flash. The event horizon is also the *static limit*.

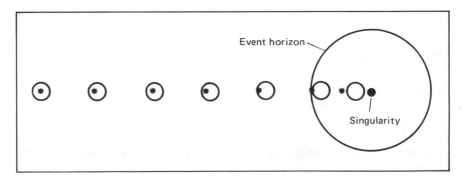

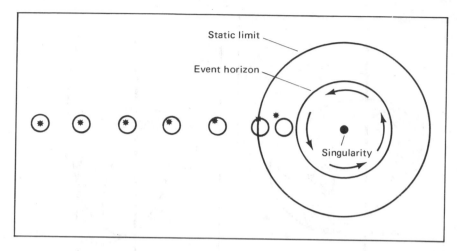

FIGURE 11-3. *Flashes of Light near a Rotating Black Hole.* Expanding spherical shells of light from flashbulbs set off near a rotating black hole are pulled *both* toward and around the hole. The combination of these two effects causes the static limit to be *above* the event horizon.

dragged both inward and around the hole. The closer the flashbulb is to the hole, the more pronounced this phenomenon is. Indeed, *above* the event horizon surrounding the black hole, there is a location where the expanding shell of light is entirely to one side of the position of the flashbulb at the instant of the flash. Therefore the static limit around a rotating black hole lies *above* the event horizon. Long before he gets to the event horizon, an astronaut in a rocketship discovers that he must be moving at the speed of light in order to remain at rest. Inside the static limit, he finds that he is dragged inward *and* around the hole, regardless of the power that his rockets have.

The fact that the static limit of a rotating black hole is located above the event horizon has some important consequences. As is the case with all black holes, once you cross the event horizon it is impossible to return to our universe. But from anywhere above the event horizon, it is always possible to get back out to our universe. Therefore, if an astronaut fell inside the static limit, he could get back out provided he did not also fall through the event horizon. In other words, there is a peculiar region of space-time around a rotating black hole where it is impossible to be at rest yet which is accessible to our universe. This region is located between the static limit and the event horizon and is called the *ergosphere.* A cross-sectional diagram of the ergosphere is shown in Figure 11-4.

One of the most interesting properties of the ergosphere was discovered in 1969 by Roger Penrose. Penrose calculated what would

175

FIGURE 11-4. *The Ergosphere* (*left*). Between the static limit and the event horizon surrounding a rotating black hole is a region of spacetime called the ergosphere. Inside the ergosphere it is impossible to be at rest, but one could go in and out of the ergosphere without ever leaving our universe.

FIGURE 11-5. *The Penrose Mechanism* (*right*). If an infalling particle breaks in two inside the ergosphere, the ejected piece can fly back out with an enormous amount of energy. The captured piece falls into the event horizon and is swallowed by the hole. Some of the hole's rotational energy is thereby transferred to the ejected piece. (Adapted from J. Wheeler.)

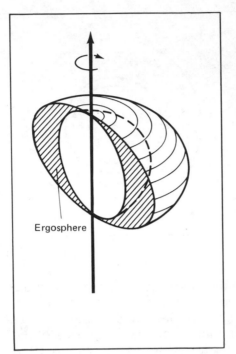

Ergosphere

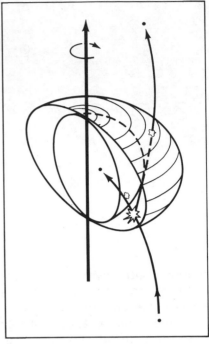

happen if an object falls into the ergosphere of a rotating black hole and breaks into two pieces. He supposed that one piece falls into the event horizon (and therefore is lost forever) while the second piece recoils back out into our universe. This situation is shown in Figure 11-5. Of course, the piece that comes out must be smaller than the object that fell in. Nevertheless, if the infalling object has exactly the right direction and speed, the energy of the ejected particle can be far *greater* than the energy of the infalling particle. And the black hole spins a little more slowly. In this way great quantities of energy can be extracted from rotating black holes. Part of the hole's rotational energy is transferred to the ejected matter by this *Penrose mechanism.*

Although astronomical implications of this phenomenon will be reserved for a later chapter, it is possible to conceive of a fictitious application of the Penrose mechanism. Suppose some advanced creatures discover a rotating black hole. Furthermore, suppose they build a city around the black hole, as shown in Figure 11-6. This city is equipped with a conveyor belt that goes down into the ergosphere, but always above the event horizon. All day long, garbage trucks collect refuse from the city and deposit it in carts on the con-

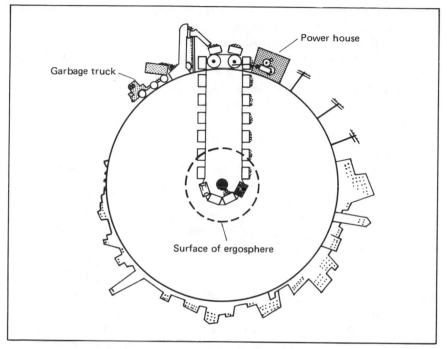

FIGURE 11-6. *The Superecological City.* When garbage from carts on a conveyor belt is dumped inside the ergosphere, the conveyor belt speeds up. If an electric generator is attached to the belt, the energy extracted from the black hole can be put to use. (Adapted from Misner, Thorne, and Wheeler.)

veyor belt. The trash is lowered into the ergosphere, at which point the garbage is dumped into the event horizon. The dumping of garbage from each cart is just like an object breaking in two pieces. Since the garbage is swallowed by the hole, each cart receives some of the hole's rotational energy. Therefore the conveyor belt receives a mighty "tug" with each dumping. The conveyor belt begins to go faster and faster. The city's inhabitants attach an electric generator to the conveyor belt, which provides them with vast quantities of power!

A somewhat less fictitious but equally fascinating application of the Penrose mechanism was discovered in the early 1970s by several astrophysicists, including Press and Teukolsky. Just as particles passing through the ergosphere can extract energy from a rotating black hole, radiation passing such a hole is also amplified. This process is known as *superradiant scattering.* To illustrate this process, imagine a black hole surrounded by a spherical mirror, as shown in Figure 11-7. Through a small hole in the mirror, shine some light in toward the black hole. As the light bounces around inside the spherical mirror, the radiation extracts more and more energy from the black hole, which gradually rotates slower and slower. As a result,

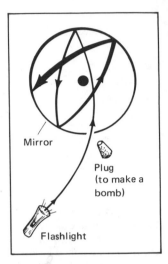

FIGURE 11-7. *Superradiant Scattering.* Light passing near a rotating black hole is amplified. If the black hole is surrounded by a mirror, the radiation can be amplified almost indefinitely. If there are no holes in the mirror, the light can be amplified so much that the mirror is blown apart, producing a *black hole bomb.*

great quantities of radiation will begin to pour back out of the little hole, thereby providing an almost unlimited source of energy. But if the little hole were plugged up just after some light was shone in, the radiation would have no way of getting back out. It would continue to bounce around inside the spherical mirror, becoming more and more powerful with each passage through the ergosphere. The resulting enormous stress on the mirror due to radiation pressure would soon become so great that the mirror would explode releasing all the trapped energy in one huge burst. This is called a *black hole bomb!*

While unusual things occur above rotating black holes, the Kerr solution has some even more unusual surprises in the convoluted space-time near the singularity. In certain respects, the geometry of rotating black holes is similar to that of charged black holes. Therefore the rest of this chapter will parallel the treatment of the Reissner-Nordstrøm solution given in Chapter 10.

Recall that the Schwarzschild black hole has a singularity surrounded by a lone event horizon. This is the simplest of all possible black holes. A nonrotating black hole is *spherically symmetric;* it is the same in all directions. But when rotation is added, the black hole is *not* the same in all directions. There are then "preferred" directions. The *axis of rotation* about which the hole spins is different from any other direction. The *equatorial plane* of the hole (which cuts through the hole perpendicular to the axis of rotation) is different from any other plane. In short, the hole has different properties in differing directions. Since the hole is rotating about an axis, the Kerr solution is called *axially symmetric.*

The most important way in which a rotating black hole varies in different directions deals with the singularity. The singularity is always a place inside a black hole where there is infinite warping of space-time. In either a Schwarzschild or a Reissner-Nordstrøm black hole, the singularity is a *point* at the center of the hole. However, when rotation is added to a black hole, the nature of the singularity changes dramatically. In a Kerr black hole, the singularity is a *ring* at the center of the hole. The *ring singularity* in a rotating black hole lies in the equatorial plane; the ring is centered about and perpendicular to the axis of rotation. In a nonrotating black hole (Schwarzschild or Reissner-Nordstrøm), anyone who travels to the center of the hole hits the singularity. However, in a rotating black hole, *only* an astronaut entering the hole along the equatorial plane hits the singularity. Only along the equatorial plane does one encounter infinite curvature of space-time. At any other angle away from the equatorial plane, an astronaut does *not* experience infinite space-

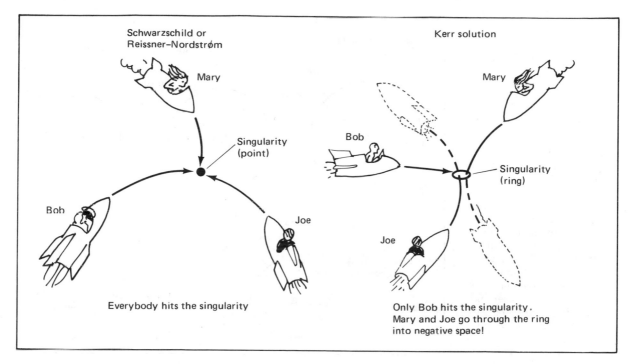

Everybody hits the singularity

Only Bob hits the singularity.
Mary and Joe go through the ring
into negative space!

FIGURE 11-8. *Singularities.* In a Schwarzschild or Reissner-Nordstrøm black hole, the singularity is a point. Anyone traveling to the center of such a hole is destroyed. In a Kerr black hole, however, the singularity is a ring through which astronauts can travel to negative or antigravity universes.

time warps. Approaching the center of a Kerr black hole at any angle out of the equatorial plane, an astronaut is *not* necessarily torn apart by infinite tidal stresses.

The ring nature of the Kerr singularity is a truly incredible property of rotating black holes. It means that an astronaut approaching the center of a Kerr black hole can pass *through* the ring without dying (see Figure 11-8). In passing through the ring singularity, the astronaut emerges into a totally new and bizarre region of space-time never before encountered. He enters *negative space.* In spite of what was said in earlier chapters, an astronaut who passes through the ring singularity is at *negative distance* from the center of the black hole. He actually can be "−10 miles" from the hole.

Some physicists dislike the idea of negative distance. In searching for an alternative interpretation of this new region of space-time, they find that it has all the properties of *antigravity!* From the other side of the ring singularity, gravity is repulsive. The black hole repels matter and light rays in this region of space-time. It is therefore sometimes called a *negative universe* or, conversely, an *antigravity universe.* This existence of antigravity universes is the most striking difference between rotating and charged black holes.

179

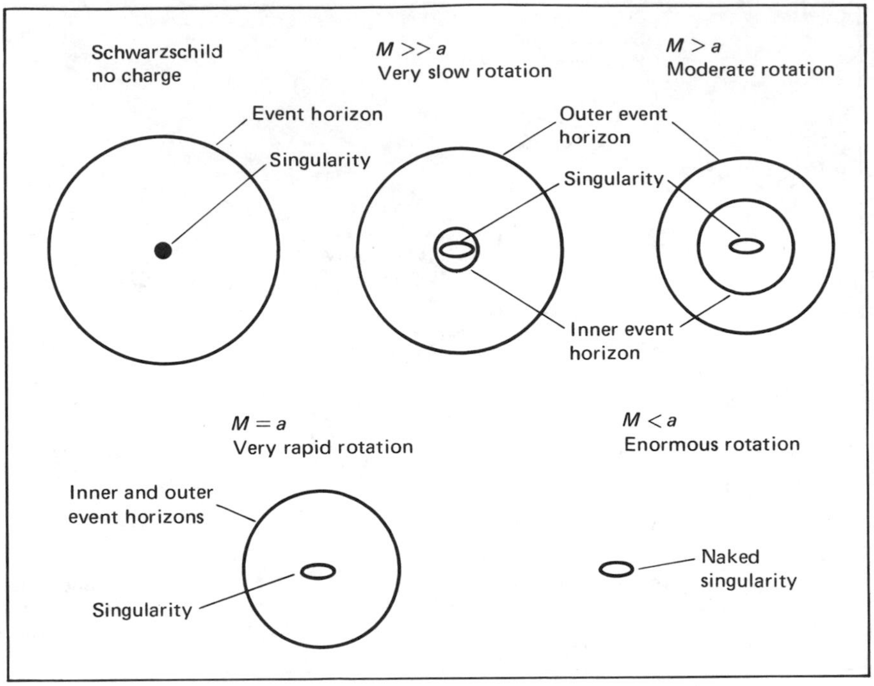

FIGURE 11-9. *Kerr Black Holes in Space.* With no rotation ($a = 0$, Schwarzschild solution) there is one event horizon surrounding a point singularity. When a small amount of rotation is added ($M>>a$), a second event horizon appears around a ring singularity. With increasing angular momentum, the two horizons get closer and closer. They merge in the extreme Kerr solution ($M=a$). For $M<a$, all horizons disappear.

In spite of the vast differences in the singularities of rotating and charged black holes, the behavior of the event horizons in both cases is quite similar. With the addition of a tiny amount of rotation ($M>>a$), a second event horizon appears very close to the singularity. With the addition of more angular momentum ($M>a$), this inner event horizon increases in size while the outer event horizon shrinks. Finally, when the hole is rotating so fast that $M=a$, the two horizons merge. This situation is often called the *extreme Kerr black hole.* If even more rotation could be added ($M<a$), all event horizons disappear and we are left with a naked ring singularity in violation of the law of cosmic censorship. A sequence of drawings showing typical locations of the event horizons for black holes having the same mass but different rates of rotation is shown in Figure 11-9.

In the previous chapter, strong arguments were presented which show that a real black hole would have very little or no charge. But simultaneously we would expect a real black hole to have large angular momentum due to its creation from a massive rotating star. But precisely how much angular momentum might a real black hole have? Is a realistic case in the range $M>>a$? Or would it be closer to the "extreme" $M=a$?

In 1974, Kip S. Thorne published realistic calculations involving black holes. He showed that, under reasonable circumstances, a black hole would be expected to be rotating at the special or *canonical* value of $a = 99.8\%\ M$. This is very rapid indeed. The techniques for drawing Penrose diagrams developed for (unrealistic) charged black holes will now pay off.

In order to discover the complete or *global* structure of space-time about a rotating black hole, it is useful to begin by drawing simplified space-time diagrams. If the singularity were a point, these diagrams would be the same as for the Reissner-Nordstrøm solution. As before, there are two event horizons that get closer and closer with increasing angular momentum. But the singularity is a ring through which astronauts can emerge into negative space. Therefore space-time diagrams must have a "left" side. The space-time diagrams must be extended to the left of the singularity to incorporate distances less than zero. In addition, the singularity in space-time diagrams of rotating black holes is drawn with dotted lines to reflect the fact that not all astronauts traveling to the center of a Kerr hole encounter infinite space-time warps, only those entering the hole along the equatorial plane. Anyone else passes through to negative space. The resulting space-time diagrams are shown in Figure 11-10 (compare with Figure 10-4).

Consider the space-time diagram for a Kerr black hole with moderate angular momentum ($M > a$). In the outside universe far above the black hole, the spacelike and timelike directions have their usual orientations. The timelike direction is vertical (parallel to the time axis) and the spacelike direction is horizontal (parallel to the space axis). But each time an event horizon is crossed, the roles of space and time are interchanged. Therefore, in between the inner and outer event horizons, the timelike direction is horizontal while the spacelike direction is vertical, as shown in Figure 11-11. Finally, in crossing through the inner event horizon, yet another role-reversal of space and time occurs. Thus, everywhere to the left of the inner event horizon in Figure 11-11, the timelike direction is again vertical while the spacelike direction is horizontal.

In order to construct Penrose diagrams of Kerr black holes, simply follow the rules set forth in the previous chapter. By way of review, these rules are as follows. There is a role-reversal of space and time crossing an event horizon. All event horizons are oriented at 45°. All outside universes must have a triangular appearance with the five infinities (see Figure 10-6). Since the Kerr solution reduces to the Schwarzschild solution when the black hole stops rotating ($a \rightarrow 0$), there must be another universe opposite ours which can be

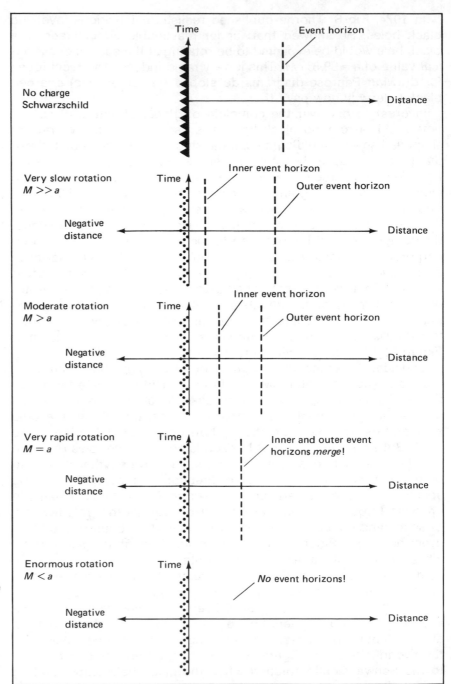

Time

Event horizon

No charge
Schwarzschild

Distance

Very slow rotation
$M \gg a$

Time

Inner event horizon

Outer event horizon

Negative
distance

Distance

Moderate rotation
$M > a$

Time

Inner event horizon

Outer event horizon

Negative
distance

Distance

Very rapid rotation
$M = a$

Time

Inner and outer event
horizons *merge*!

Negative
distance

Distance

Enormous rotation
$M < a$

Time

No event horizons!

Negative
distance

Distance

FIGURE 11-10. *Space-Time Diagrams of Kerr Black Holes.* This sequence of diagrams shows the simplified structure of space-time for black holes with the same mass (M) but different rates of rotation (a). The singularity is indicated with dotted lines because it is possible to pass through to regions of negative distance.

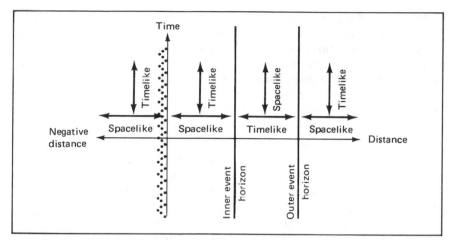

FIGURE 11-11. *A Space-Time Diagram for a Kerr Black Hole* (*M* > *a*). The orientations of spacelike and timelike directions are shown in this diagram for a nonextreme Kerr black hole. Each time an event horizon is crossed, the roles of space and time are interchanged.

reached only along spacelike paths. Finally, since there are two event horizons and thus two role-reversals of space and time in going from the outside universe to the singularity, the singularity must be timelike. It must be oriented vertically in the Penrose diagram.

To begin constructing a conformal map of all space-time, imagine an astronaut who blasts off from the earth and plunges toward a rotating black hole. He passes through the outer event horizon and later falls through the inner event horizon. As shown in Figure 11-12, our universe has the usual triangular appearance and the event horizons are oriented at 45°.

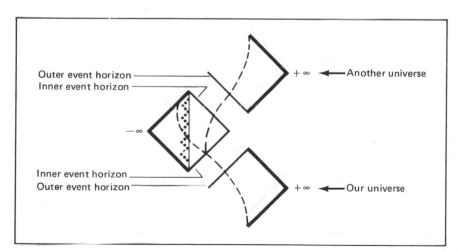

FIGURE 11-12. *A Piece of a Penrose Diagram.* The global structure of space-time can be deduced by following an astronaut into a rotating black hole. The trip shown here is for an astronaut who leaves from the earth in our universe. (Compare with Figure 10-8.)

Upon crossing the inner event horizon, the astronaut has several choices. If he (unfortunately!) happens to be in the equatorial plane, he can hit the singularity, which must be vertical (timelike) in the Penrose diagram. But if he approaches the center of the hole at an oblique angle, he passes through the ring singularity into the negative universe. Since he does not get torn apart in passing through the ring, the singularity is drawn with dotted lines. As usual, the negative universe has a triangular appearance in the conformal map.

Since the singularity is timelike and therefore vertical, the astronaut could have completely avoided both of the above choices by simply directing his rocketship outward. In leaving the black hole, he passes out through the inner event horizon and then later out through the outer event horizon. He emerges into a future universe. He can stay in this new universe visiting planets, or he can turn around and plunge back into the hole, thereby traveling to more and more future universes.

To get the rest of the Penrose diagram, realize that if the black hole stopped rotating, everything must reduce to the Schwarzschild geometry (see Figure 9-18). This means that there must be another universe opposite ours which can be reached only along spacelike (forbidden) paths. Consider, therefore, the trip of an alien astronaut who leaves a planet in this "other" universe and plunges into the rotating black hole in a flying saucer. The choices in making the trip are the same as the astronaut from earth had. As shown in Figure 11-13, the alien can hit the singularity by traveling in the equatorial plane or he can pass through to the negative universe by approach-

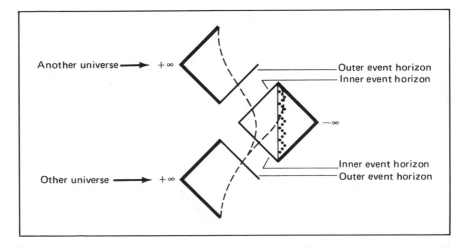

FIGURE 11-13. *Another Piece of a Penrose Diagram.* Another section of the Penrose diagram can be obtained by following an alien astronaut (appropriately equipped with flying saucer) from another universe in the alien's travels through the rotating black hole. (Compare with Figure 10-9.)

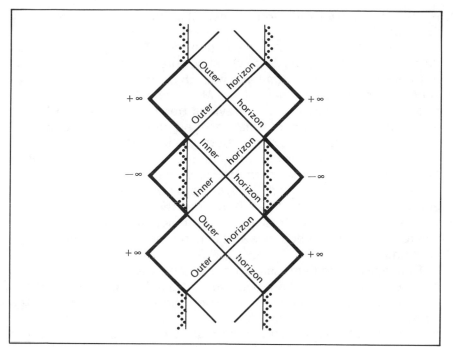

FIGURE 11-14. *The Entire Penrose Diagram of a Kerr Black Hole (M > a).* The full Penrose diagram is obtained by assembling the pieces from Figures 11-12 and 11-13. The diagram repeats over and over infinitely into the future and past like a long string of paper dolls. (Compare with Figure 10-10.)

ing the ring singularity obliquely. Alternatively, he could wander in and out of the event horizons, thereby going to all sorts of future universes.

Finally, to obtain the complete picture, the two pieces (Figures 11-12 and 11-13) are put together. The resulting Penrose diagram for a Kerr black hole is shown in Figure 11-14. Since an astronaut could go indefinitely in and out of the event horizons from one universe to the next, the diagram must go on forever into the future and the past.

Notice that the Penrose diagram for a Kerr black hole $(M > a)$ looks very similar to the Penrose diagram for a Reissner-Nordstrøm black hole $(M > |Q|)$ shown in Figure 10-10. Only one important difference stands out. In a charged black hole, the singularity is a point and everyone approaching the center of such a hole experiences infinitely warped space-time, which completely cuts off any hope of ever traveling into negative space. With a rotating black hole, however, it is possible to pass through the ring singularity into negative space. Only in the equatorial plane does an ill-fated astronaut get torn apart. In a Penrose diagram of a Kerr black hole, the singularity is therefore shown with dotted lines. It is the portal to antigravity universes.

In the Reissner-Nordstrøm solution, the three possible cases $(M > |Q|, M = |Q|, \text{ and } M < |Q|)$ had three very different-appearing Penrose diagrams. Similarly, with the Kerr solution, the corresponding three possibilities $(M > a, M = a, \text{ and } M < a)$ have very different Penrose diagrams. The previous treatment leading to Figure 11-14 was for the case of small or moderate angular momentum $(M > a)$. To examine the extreme Kerr geometry $(M = a)$, return once more to a simplified space-time diagram. In an extreme Kerr black hole, the inner and outer event horizons merge. As they do, the intermediate region between the two horizons disappears. Thus, as shown in Figure 11-15, there is no net reversal of spacelike and timelike directions in crossing the (double) event horizon. Everywhere the timelike direction is vertical and the spacelike direction is horizontal.

To construct the Penrose diagram for an extreme Kerr black hole, again consider an astronaut blasting off from the earth and plunging into the hole. In crossing the lone event horizon he encounters the singularity. But since there was no net change in spacelike and timelike directions, the singularity must be timelike and vertical in the Penrose diagram. The astronaut now faces several possibilities. By traveling in the equatorial plane he can strike the singularity, where life becomes unpleasant. Alternatively, the astronaut could approach the center of the black hole at an angle out of the equatorial plane. In this case he passes through the ring singularity and

FIGURE 11-15. *A Space-Time Diagram for an Extreme Kerr Black Hole* $(M = a)$. When a black hole is rotating so fast that $M = a$, the inner and outer event horizons merge. The region between the horizons vanishes and there is no net change in spacelike and timelike directions after crossing the (double) horizon.

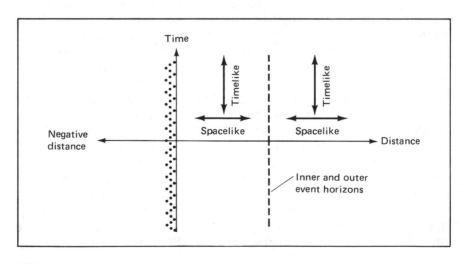

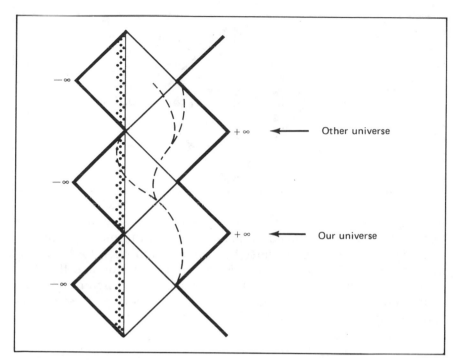

FIGURE 11-16. *The Penrose Diagram for an Extreme Kerr Black Hole (M = a).* A conformal map of an extreme Kerr black hole can be deduced by following the possible paths of an astronaut. As usual, the diagram is repeated over and over infinitely into the future and the past. (Compare with Figure 10-13.)

emerges into an antigravity universe that has the usual triangular appearance. The third choice is to avoid the center of the black hole completely, turn around and go back out through the event horizon into an ordinary future universe, as shown in Figure 11-16. He can then either remain in this new universe visiting planets or return to the black hole and encounter all the same possibilities once again. The Penrose diagram therefore repeats over and over infinitely far into the past and the future.

Notice once again that the Penrose diagram for the extreme Kerr solution is very similar to the Penrose diagram for the extreme Reissner-Nordstrøm solution. The only basic difference is that it is possible to go through the Kerr singularity into antigravity universes.

Finally, if the black hole is spinning so fast that $M < a$, the event horizons vanish, exposing a naked singularity to the outside universe. Unlike the case with the Reissner-Nordstrøm naked singularity, however, an astronaut can pass through the ring singularity and emerge in an antigravity universe. The resulting Penrose diagram, therefore, has the very simple appearance shown in Figure 11-17. In addition, an astronomer can see light from the antigravity universe

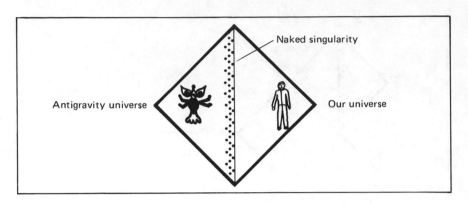

FIGURE 11-17. *The Naked Kerr Singularity.* If a black hole is rotating so rapidly that $M < a$, all event horizons disappear, thereby exposing a naked singularity. Astronauts can travel through the ring singularity separating our universe from the antigravity universe.

passing through the ring singularity. And, conversely, the alien astronomer in the antigravity universe can see light from our universe.

Since real black holes must be rotating and therefore are described by the Kerr geometry, it is instructive to examine the Kerr solution in greater detail. Specifically, in the next chapter considerable attention will be devoted to what astronomers and astronauts view as they observe or explore rotating black holes.

12 THE GEOMETRY OF THE KERR SOLUTION

Theoretical astrophysicists working with mathematics often face many choices. Specifically, they can make life easy or difficult for themselves by setting up their equations in either convenient or cumbersome fashion. This is especially true when examining the geometry of a rotating black hole.

In describing the geometry of space about a Kerr black hole, physicists have several choices for denoting positions and locations around the hole. These possibilities amount to choosing a *coordinate system,* which is simply a grid with which to cover all space. For example, the physicist could choose rectangular coordinates. Rectangular coordinates, shown at the left of Figure 12-1, look like ordinary graph paper. A position in rectangular coordinates is given by the distances in the up-down and left-right directions.

Physicists would be very foolish to choose rectangular coordi-

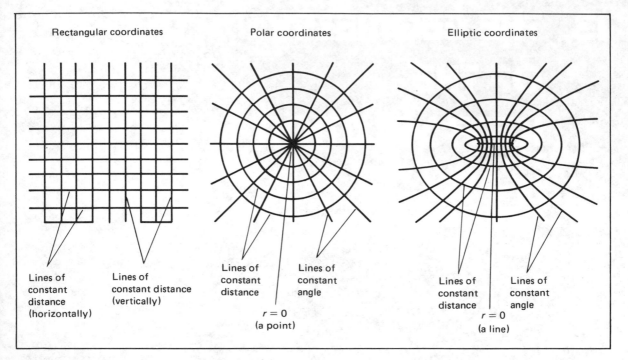

Rectangular coordinates   Polar coordinates   Elliptic coordinates

Lines of constant distance (horizontally)

Lines of constant distance (vertically)

Lines of constant distance

Lines of constant angle

$r = 0$
(a point)

Lines of constant distance

Lines of constant angle

$r = 0$
(a line)

FIGURE 12-1. *Various Coordinate Systems.* A coordinate system is simply a grid used to denote locations in space. The convenient choice for rotating black holes is elliptic coordinates rotated about the axis of symmetry. This coordinate system properly reflects the geometry of the Kerr solution.

nates for describing space about a black hole. Rectangular coordinates are useful for objects having right angles, but black holes do not look like bricks. Rectangular coordinates do not reflect the symmetries of black holes and therefore the physicists' equations would be very messy.

A second possible choice involves the use of polar coordinates. As shown in the middle of Figure 12-1, locations in polar coordinates are measured from a point. A location in polar coordinates is given by the distance from the point and an angle.

Polar coordinates (extended to three dimensions) are ideal for any situation that is spherically symmetric. Schwarzschild and Reissner-Nordstrøm black holes are spherically symmetric. The polar coordinate system is therefore ideally suited for describing space in the Schwarzschild and Reissner-Nordstrøm solutions, and the equations written in spherical polar coordinates take on a very simple form.

Although spherical polar coordinates are great for a Schwarzschild or Reissner-Nordstrøm black hole, they are not as convenient for the Kerr solution. A rotating black hole is *not* spherically symme-

190

tric. There is a preferred direction, the axis of rotation, about which the hole spins. In working with the Kerr solution, physicists need to choose a coordinate system that properly reflects the geometry of a rotating black hole or else their equations will be unnecessarily complicated.

There is another coordinate system that can be readily adapted to the Kerr solution. In two dimensions it is called elliptic coordinates; it is shown at the right of Figure 12-1. In essence, locations are given by the distance from a line and an angle. Curves of constant distance from the line are ellipses and curves of constant angle are hyperbolas. Elliptic coordinates may be thought of as polar coordinates with the point at the center stretched into a line.

To make a three-dimensional coordinate system suitable for the Kerr solution, imagine the elliptic coordinates rotated about the axis of symmetry. The ellipses become ellipsoids and the hyperbolas become hyperboloids. The end points of the line at the center of the coordinate system trace out a ring. The resulting three-dimensional system is called *oblate spheroidal coordinates* and is shown in Figure 12-2.

Oblate spheroidal coordinates are perfect for the Kerr solution. The system is axially symmetric just as a rotating black hole is axially symmetric. The center of the system is a ring just as the Kerr singularity is a ring. Wise physicists therefore use oblate spheroidal coordinates in their calculations. Although no computations will be presented here, it nevertheless is important to note the basic properties of these coordinates. Looking down the axis of rotation, we see

FIGURE 12-2. *Oblate Spheroidal Coordinates.* Oblate spheroidal coordinates are obtained by rotating elliptic coordinates about the axis of symmetry. The center of the coordinates is a ring. This axially symmetric system is ideally suited for the Kerr solution because the Kerr singularity is a ring.

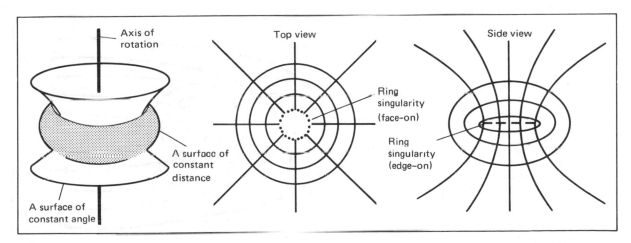

that the coordinate system (or a Kerr black hole) has a circular appearance. However, looking in along the equatorial plane, we see that the coordinate system (or a Kerr black hole) has an elliptical appearance, as shown in Figure 12-2.

When introducing the Schwarzschild black hole in Chapter 8, it was very instructive to follow the paths of light rays as shown, for example, in Figure 8-1. As beams of light passed near the black hole, they were deflected by the curvature of space-time. It was found that beams of light approaching the black hole at exactly the right distance would go into *circular orbit* about the hole. This phenomenon gives rise to the *photon sphere,* a single sphere of unstable circular light orbits. For convenience, a diagram of light orbits for a Schwarzschild black hole is shown in Figure 12-3.

It is extremely important to emphasize the fact that in a Schwarzschild black hole there is *one* photon sphere. There is only one distance above the event horizon at which circular light orbits can exist. In addition, the light rays in the photon sphere are orbiting the hole at every possible angle, either clockwise or counterclockwise. For an incoming light ray approaching the hole at exactly the right distance to be captured into the photon sphere, it does not matter exactly where the light ray originated. The angles at which light rays approach the hole are irrelevant. Indeed, since the hole is spherically symmetric, it does not have any "top" or "bottom"; there is no "left side" or "right side." The distance from the hole or *impact parameter* of the light ray is the only thing that matters. With the proper impact parameters, *all* light rays go into the *same* photon sphere, regardless of what star or direction they came from.

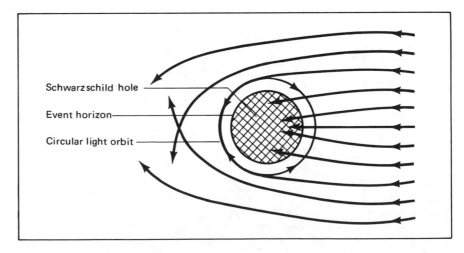

FIGURE 12-3. *Light Orbits about a Schwarzschild Black Hole.* A nonrotating black hole is surrounded by a sphere of unstable circular light orbits. *Any* beam of light approaching the hole at exactly the appropriate distance can be captured into circular orbit in the photon sphere.

192

With a rotating black hole, the situation is very different. In the case of a Kerr black hole, the axis of rotation about which the hole spins defines a specific direction in space. Space-time is curved differently at different angles from the axis of rotation. The geometry of space-time is axially symmetric, not spherically symmetric. These complications result in major changes in the locations of circular light orbits.

To begin understanding light orbits about a Kerr black hole, imagine looking straight down the axis of rotation at light rays approaching the hole in the equatorial plane. As shown in Figure 12-4, light rays passing far from the hole (in other words, with large impact parameters) are deflected only slightly. With exactly the right impact parameter, light beams can again go into circular orbit about the hole. But now there are *two* possibilities. Approaching the hole from one side, light rays can be captured into an unstable circular orbit provided the light revolves about the hole opposite to the direction in which the hole is rotating. This *counterrotating circular orbit* is at a greater distance from the hole than the photon sphere of the Schwarzschild case.

Approaching the hole from the other side, light rays can again be captured into unstable circular orbit. Such light rays, however, are revolving about the hole in the same direction the hole is rotating. This *corotating circular orbit* is much closer to the hole. The corotating orbit is at a smaller distance from the hole than the photon sphere of the Schwarzschild case.

By examining light rays in the equatorial plane, we see that there are two circular orbits, an inner one which is corotating with the

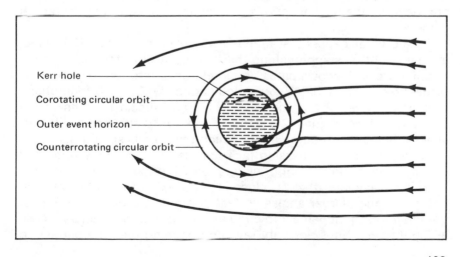

Kerr hole

Corotating circular orbit

Outer event horizon

Counterrotating circular orbit

FIGURE 12-4. *Light Orbits about a Kerr Black Hole* (*in equatorial plane*). Light rays passing far from the rotating black hole are deflected through small angles. A light ray approaching the hole with the required impact parameter can go into circular orbit about the hole. In the equatorial plane, however, there are *two* unstable circular light orbits. The outer orbit is for counterrotating light rays while the inner orbit is for corotating light rays.

193

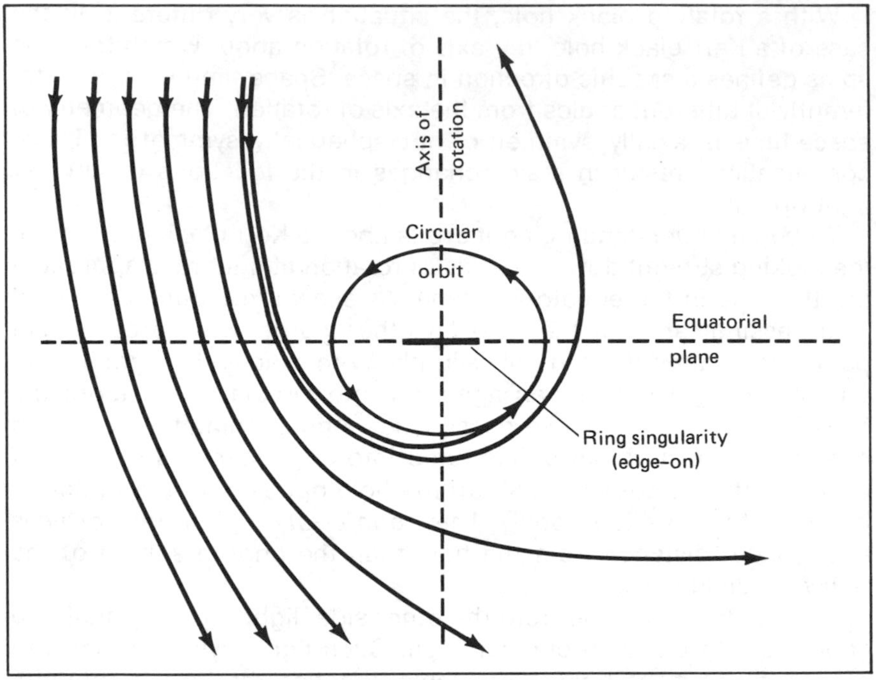

FIGURE 12-5. *Light Orbits about a Kerr Black Hole (parallel to axis of rotation).* Light rays passing far from the rotating black hole are deflected through small angles. There is only one possible circular light orbit for a light ray approaching the hole parallel to the axis of rotation. (This diagram is for the extreme Kerr solution where $M = a$.)

hole and an outer one which is counterrotating. In essence, when angular momentum is added to a Schwarzschild black hole, the photon sphere "splits" in two. In between the corotating and counterrotating orbits in the equatorial plane there are many possible unstable circular light orbits. These orbits, however, are for light rays approaching the hole at various directions inclined to the equatorial plane.

To see what happens at angles away from the equatorial plane, imagine light rays approaching the hole parallel to the axis of rotation. Figure 12-5 shows paths of such light rays for an extreme Kerr black hole ($M = a$) based on calculations by C. T. Cunningham. Whereas Figure 12-4 is the "top view" and displays orbits in the equatorial plane, Figure 12-5 is the "side view" displaying orbits parallel to the axis of rotation about which the hole is spinning.

As usual, light rays passing far from the black hole are deflected through small angles. Light rays approaching the hole with smaller impact parameters (that is, closer to the axis of rotation) are deflected through larger angles. In fact, there is *one* particular impact parameter for light paths into circular orbit about the hole, as shown in Figure 12-5. Thus, for light rays approaching the hole parallel to

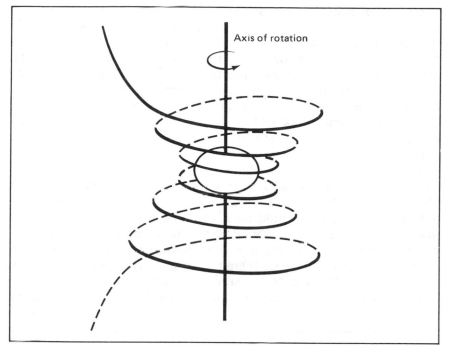

Axis of rotation

FIGURE 12-6. *A Light Ray Passing a Kerr Black Hole.* As a light ray passes near a rotating black hole, the path is pulled around the hole by the dragging of space-time. The orbits shown in Figure 12-5 (and all similar diagrams) should therefore be rotated about the hole's axis to obtain the real paths in three-dimensional space.

the axis of rotation, there is *one* unstable circular orbit. The distance of this orbit from the black hole is in between the distances for the counterrotating and corotating orbits in the equatorial plane. If you are puzzled by the fact that the "circular" orbit in Figure 12-5 is an ellipse, recall that oblate spheroidal coordinates are used. In the "side view" of these coordinates (see Figure 12-2), a curve of constant distance from the ring singularity is an ellipse.

In a certain sense, Figure 12-5 is an oversimplification. Space-time surrounding a rotating black hole is dragged around with the hole. Although Figure 12-5 properly displays the distances of incoming light rays from the black hole, that is all the diagram shows. In reality, as a light ray approaches a rotating black hole, it begins to spiral around due to the dragging of inertial frames. Figure 12-6 shows this phenomenon for one particular light ray. The complete path in three-dimensional space is a complicated helix. Therefore, to obtain the full picture of what happens to incoming light rays, Figure 12-5 (and any similar diagram) should be rotated about the axis of the black hole. Nevertheless, diagrams like Figure 12-5 perform an important function by showing the distance (only) of a light ray from the hole.

By way of summary, we see that there are a variety of unstable circular orbits around a rotating black hole. The orbit farthest from the hole is the counterrotating orbit in the equatorial plane. The one nearest the hole is the corotating orbit, also in the equatorial plane. In between these extremes there is a range of possible orbits for light rays approaching the hole at different angles. For any given angle there are both corotating and counterrotating orbits, except for a beam of light parallel to the axis of rotation. For a light ray approaching the hole parallel to the axis of rotation, there is only one circular orbit.

For a slowly rotating black hole, the range of circular orbits is not large. All the possible orbits are crowded together at a distance above the outer event horizon near the location where the Schwarzschild photon sphere would be *if* the hole were not rotating. For more rapidly rotating black holes, the distance between the corotating and counterrotating orbits in the equatorial plane is larger. The range of circular orbits is therefore larger. The largest possible range occurs with the extreme Kerr black hole ($M = a$).

To show the range of circular light orbits about a rotating black

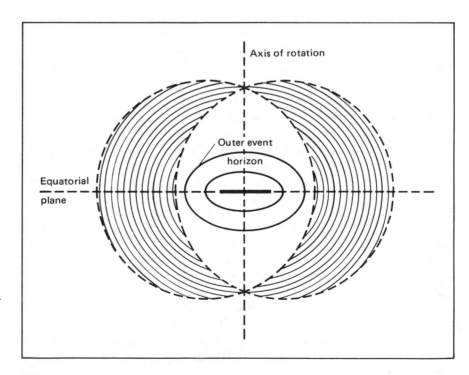

FIGURE 12-7. *The Range of Circular Light Orbits above a Rapidly Rotating Black Hole.* All possible circular light orbits above a Kerr black hole ($a = 90\% \, M$) lie inside the boundaries shown here. A particular light ray in circular orbit precesses in a very complicated fashion on the surface of an ellipsoid inside the boundaries.

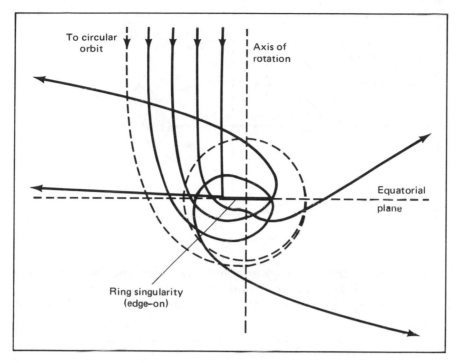

To circular
orbit

Axis of
rotation

Equatorial
plane

Ring singularity
(edge–on)

FIGURE 12-8. *Light Paths into a Kerr Black Hole.* Light rays aimed at a rotating black hole with impact parameters less than that for circular light orbits go into the hole. The paths followed by light rays deep inside the hole reveal the singularity to be repulsive. Near the singularity, light rays experience antigravity. (This diagram is for the extreme Kerr solution where $M = a$.)

hole, it is most convenient to draw the boundaries or *envelope* of all such orbits. Figure 12-7 shows the envelope of all possible circular orbits above a rapidly rotating Kerr hole ($a = 90\% \ M$). In general, a particular light ray in orbit about the hole precesses in a very complicated fashion along the surface of an elliptical ring inside the boundaries. If angular momentum were removed from the black hole, as the hole rotated slower and slower the volume inside the boundaries would decrease. If the black hole stopped rotating, the entire envelope would reduce to the photon sphere of a Schwarzschild black hole.

The previous treatment deals only with what happens *above* a Kerr black hole. To probe the geometry of space-time *inside* a Kerr black hole, imagine sending light rays in at impact parameters *less* than is required for circular light orbits. Figure 12-8 shows light rays approaching a Kerr black hole parallel to the axis of rotation at impact parameters smaller than that needed for insertion into circular orbit. Figure 12-8 is simply a continuation of Figure 12-5 and is again based on calculations by C. T. Cunningham. The important fact to notice from the paths of these light rays is that near the

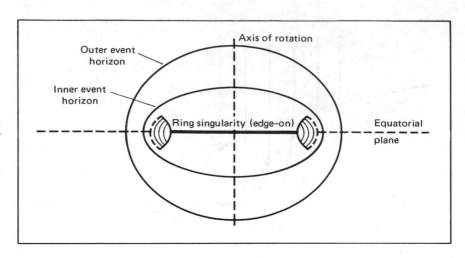

FIGURE 12-9. *The Range of Circular Light Orbits inside a Rapidly Rotating Black Hole.* Inside the inner event horizon there is a region where the attractive gravitational field of the hole is balanced by the repulsive field of the singularity. In this region both stable and unstable circular light orbits are permitted. (This diagram is for the case $a = 90\%\ M$.)

center of the black hole they are bent *away* from the singularity. Although gravity far from the center of a Kerr black hole is attractive and pulls things inward, near the singularity gravity is repulsive and

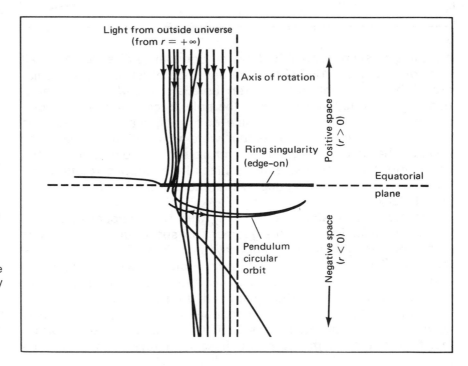

FIGURE 12-10. *Light Paths through the Ring Singularity.* The upper half of this diagram is in positive space (where the light rays come from) and the lower half is in negative space (where the light rays go to). The light rays are deflected away from the ring singularity due to the repulsive properties of gravity at the singularity. Some light rays can go into circular orbits in negative space. (This diagram is for the extreme Kerr solution where $M = a$.)

tries to push things out! Indeed, the light ray aimed directly at the ring is severely deflected; the beam of light literally bounces right back out of the hole. This repulsive nature of the Kerr singularity means that at a certain distance from the center of the hole the attractive and repulsive aspects of gravity are balanced. Therefore, in this balanced region it is again possible to have circular light orbits! Figure 12-9 shows the boundaries of all possible circular light orbits deep inside the inner event horizon. Unlike the outer orbits above the black hole, both unstable and stable orbits are possible in this inner region. The singularity of a Kerr black hole is therefore surrounded by light rays.

To explore the innermost portions of a Kerr black hole, imagine sending in light rays parallel to and *very* near the axis of rotation. Specifically, suppose these light rays have impact parameters *less* than that needed to hit the ring singularity. Light rays traveling along or very near the axis of rotation will therefore pass through the singularity into negative space. In order to draw the complete paths of these rays, the diagram must therefore include negative space. In Figures 12-5 and 12-8, the light rays never go through the singularity. Thus they always remain in positive space; their distances from the singularity are always given by positive numbers. But when objects go into negative space, their distances from the singularity are given by negative numbers. Figure 12-10 copes with this difficulty by simply having the upper half of the diagram in positive space while the lower half is in negative space. In Figure 12-10, therefore, a light ray traveling on or very near the axis of rotation follows a very straight path from positive space through the center of the ring into negative space.

In examining light rays passing through the singularity, we first of all notice that the beams are deflected *away* from the edges of the ring. This is again due to the repulsive nature of gravity along the singularity. But there is a surprise. Notice that there is one light ray shown in Figure 12-10 which goes through the singularity and bounces back and forth along a piece of an ellipse in negative space. Now recall that ellipses are curves of constant distance from the singularity (remember that Figure 12-2 showed oblate spheroidal coordinates). Thus, this light ray is always at a constant distance from the singularity in negative space. It therefore is in circular orbit! Since it bounces back and forth, its path is called a *pendulum circular orbit.* A typical pendulum circular orbit in negative space is shown diagrammatically in Figure 12-11.

Although Figure 12-10 shows only one particular light ray going

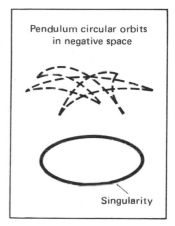

FIGURE 12-11. *Pendulum Circular Orbits in Negative Space.* Light rays passing through the singularity with exactly the appropriate impact parameter go into circular orbit above the singularity in negative space. These orbits are called pendulum because the light rays bounce around on a surface of constant distance (an ellipsoid) at a negative distance from the singularity.

199

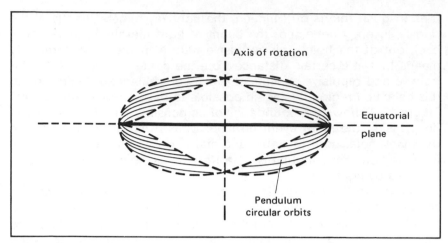

FIGURE 12-12. *The Range of Pendulum Circular Light Orbits in Negative Space* ($r < 0$). All possible pendulum circular orbits near the singularity of a Kerr black hole ($a = 90\%$ *M*) lie inside the boundaries shown in the figure. Inside this region in negative space, light rays bounce around on ellipsoidal surfaces.

into a pendulum circular orbit, there is a range of impact parameters for light rays nearly parallel to the rotation axis which can be captured into these peculiar orbits. As a result, there is a range of pendulum circular orbits in the negative universe. Figure 12-12 shows the boundaries of all possible pendulum circular orbits for a rapidly rotating black hole. Note that Figure 12-12 is a drawing entirely in negative space while the corresponding Figures 12-7 and 12-9 are entirely in positive space. All pendulum circular orbits are unstable.

To complete this treatment of light rays, notice from Figure 12-10 that it is possible for a light ray passing near the inner edge of the ring to go into negative space and bounce out again. The fact that light rays can momentarily dip into negative space and bounce back out will be important when we consider the appearance of a Kerr hole as seen by a distant astronomer.

Finally, consider light rays approaching the Kerr singularity from the negative universe. Light rays traveling on or very near the axis of rotation pass directly through the ring singularity into positive space. However, as shown in Figure 12-13, *all* light rays traveling toward the black hole with large impact parameters are *repelled* by the hole. As seen from the negative universe, the hole is a source of antigravity. It repels everything and attracts nothing. The negative universe is therefore sometimes called an antigravity universe.

Now that all the various paths and trajectories of light rays around a Kerr black hole have been presented in detail, it is possible to understand what a rotating black hole looks like to a distant astronomer or adventurous astronaut. First of all, consider an as-

200

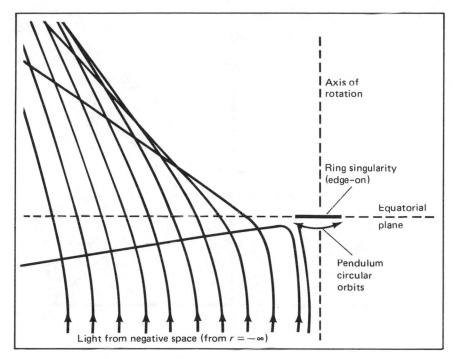

Axis of
rotation

Ring singularity
(edge-on)

Equatorial
plane

Pendulum
circular
orbits

Light from negative space (from $r = -\infty$)

FIGURE 12-13. *Light Rays from Negative Space.* Light rays approaching a rotating black hole from negative space are repelled by the hole. In negative space, a rotating black hole is a source of antigravity. (This diagram is for the extreme Kerr solution where $M = a$.)

tronomer in our universe who looks at a Kerr black hole. Since the hole is axially symmetric, he will see a different view at different angles from the hole's axis of rotation. For future reference, Figure 12-14 defines the *azimuthal angle* θ. When $\theta = 0$, the distant astronomer is looking straight down the axis of rotation. When $\theta = 90°$, he is looking in along the equatorial plane.

Imagine an astronomer with a very powerful telescope who examines the center of a rotating black hole. The astronomer is far from the hole where space-time is flat and focuses his telescope directly on the singularity. Figure 12-15, based on calculations by Cunningham, displays what the astronomer sees at various angles for the case of an extreme Kerr hole ($M - a$). Looking straight down the axis of rotation ($\theta = 0$), he sees a circular region filled with light pouring through the ring singularity from the negative universe. If the singularity itself emits light (and it *does* for reasons to be discussed in a later chapter), light from the singularity is seen in a ring surrounding the disk of light from negative space. In between the disk of light from negative space and the ring of light from the singularity, there is a region filled with light from positive space which

201

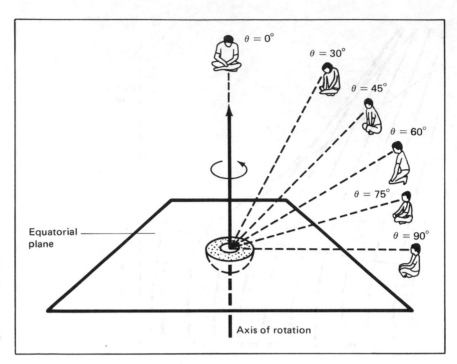

$\theta = 0°$

$\theta = 30°$

$\theta = 45°$

$\theta = 60°$

$\theta = 75°$

$\theta = 90°$

Equatorial plane

Axis of rotation

FIGURE 12-14. *The Azimuthal Angle θ*. A Kerr black hole has a different appearance at different angles. The azimuthal angle θ is useful in denoting the precise direction at which the black hole is viewed.

dips into negative space and bounces back out. Light from an earlier (positive space) universe which falls on the hole near the inner edge of the ring singularity experiences a strong antigravity field. This light is therefore repelled by the singularity and is ejected back out into the positive space of our universe. It should again be emphasized that light rays can escape from a Kerr hole because we are discussing the highly theoretical, idealized case. This complete Kerr solution really involves black and white holes.

At an angle away from the axis of rotation, the disk of light from negative space is smaller and elliptical. At greater angles, the region of light from the negative universe decreases in size and becomes even more elliptical. In addition, the ring of light from the singularity also becomes more elliptical in appearance as the Kerr singularity is seen more and more edge-on. As before, the region between the light from negative space and the light from the singularity contains rays from positive space that momentarily dip into the negative universe and are repelled back out.

The treatment just discussed is restricted to the appearance of the singularity alone. If the astronomer viewing the black hole uses a

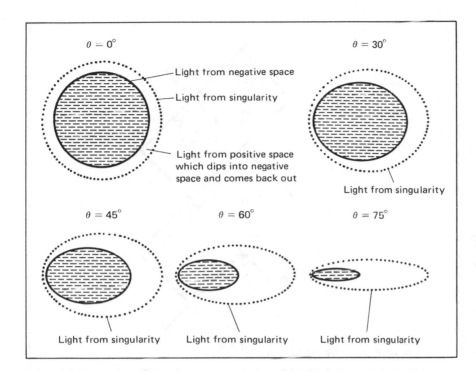

FIGURE 12-15. *The Appearance of the Singularity.* The appearance of the singularity of an extreme Kerr black hole ($M=a$) viewed at various angles is shown in this sequence of diagrams. Light from negative space pours through the center of the ring singularity. Light from the singularity itself is shown with dotted lines.

very-wide-angle eyepiece in his telescope, he will see things at large distances from the singularity. To understand the overall appearance of a rotating black hole, it is necessary to examine Penrose diagrams.

Consider the Penrose diagram for an extreme Kerr hole ($M=a$), as shown in Figure 12-16. Recall that light rays always travel along 45° lines in all such space-time diagrams. This Penrose diagram displays the paths of typical light rays that can reach the astronomer in our universe (Universe 3). First of all, he gets light from $\mathscr{I}^-$ in the negative universe (Universe 2). This light comes through the very center of the ring singularity. He also receives light from the singularity that bounds Universe 2 and thereby separates positive space (to the right) from negative space (to the left). The appearance of the light from Universe 2 and the singularity was treated in Figure 12-15. Outside the light from the singularity, however, the astronomer sees rays coming from two additional sources.

Stars and galaxies in our universe and, presumably, in other universes are shining light in all directions. Some of this light falls onto the rotating black hole. As this light passes inside the ergosphere of

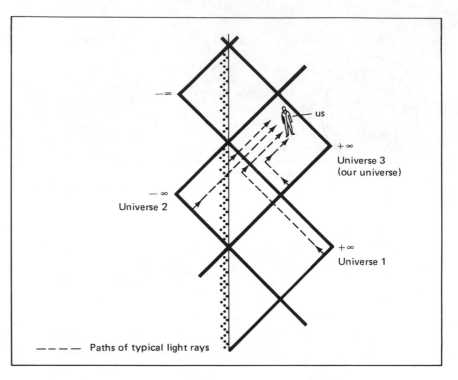

FIGURE 12-16. *The Penrose Diagram for an Extreme Kerr Black Hole (M = a)*. An astronomer ("us") in our universe receives light from a variety of locations when viewing a rotating black hole. Light from negative space (Universe 2) and the associated singularity reaches the astronomer in our universe (Universe 3). Also, light from the previous universe (Universe 1) and the early part of our universe are reflected inside the hole toward the astronomer.

the hole, it is spun around many times. Roughly speaking, some of this light experiences "centrifugal forces" that repel the beams back out into space. In other words, light rays from $\mathscr{I}^-$ in our universe and $\mathscr{I}^-$ in the previous universe (Universe 1) can be reflected back out into positive space. The distant astronomer can therefore see light from Universe 1 and light from the early history of our universe!

Figure 12-17A through C has views showing the full appearance of an extreme Kerr hole as seen by a distant astronomer in our universe. In all cases, the appearance of the singularity is simply copied from Figure 12-15. In all cases, the central part of the hole is surrounded by a large circular region filled with light from Universe 1. This light was reflected toward the astronomer from deep inside the hole. Outside of this circular region, the astronomer sees light from objects in his own universe. The astronomer examining a rotating black hole can therefore see what is going on in a negative universe and what went on in the previous positive universe. In addition, the light from Universe 3 seen near the hole actually originated early in our own universe (it came from $\mathscr{I}^-$ in Universe 3). In principle, therefore, the astronomer should be able to see what was going on bil-

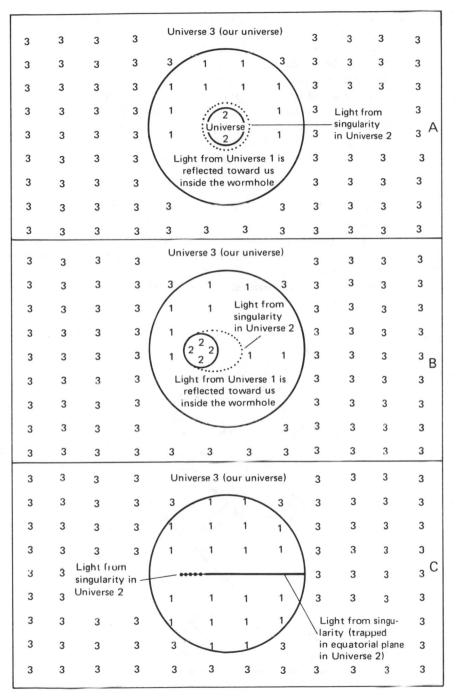

FIGURE 12-17A through C.
A: *The Appearance of an Extreme Kerr Black Hole (M = a) at θ = 0°.* Looking straight down the axis of rotation, a distant astronomer sees light from a negative-space universe and the previous positive-space universe. He also sees light from the earliest times of his own universe.

B: *The Appearance of an Extreme Kerr Black Hole (M = a) at θ = 45°.* At an angle between the axis of rotation and the equatorial plane, the appearance of the black hole is essentially the same as in Figure 12-17A. However, since the hole is rotating, the apparent location of the singularity is offset from the center of the field of view.

C: *The Appearance of an Extreme Kerr Black Hole (M = a) at θ = 90°.* Looking in along the equatorial plane, the astronomer sees the singularity edge-on. Light orbiting the singularity in the equatorial plane can spiral out to the distant astronomer.

205

lions of years ago! In principle, he should be able to see the formation of the earth, or dinosaurs, or prehistoric human beings, depending on precisely where he looks.

The appearance of the black hole at various angles has the same general features. However, at angles away from the axis of rotation, the apparent location of the singularity is offset to one side because the hole is rotating. When the astronomer views the hole along the equatorial plane ($\theta = 90°$), he sees the singularity edge-on. But unlike previous views, light orbiting the singularity in the equatorial plane can spiral out into space and be seen by the distant astronomer.

When a black hole is rotating at a slower rate than the extreme speed, the views at angles away from the equatorial plane are essentially the same as for the extreme Kerr case. But as viewed in the equatorial plane ($\theta = 90°$) some new features appear. To understand the origin of these new features, it is necessary to examine the appropriate Penrose diagram. Figure 12-18 shows the Penrose diagram for a Kerr hole with $M > a$.

The astronomer in our universe (Universe 3) still receives light directly from Universe 2 and the singularity bounding Universe 2. He

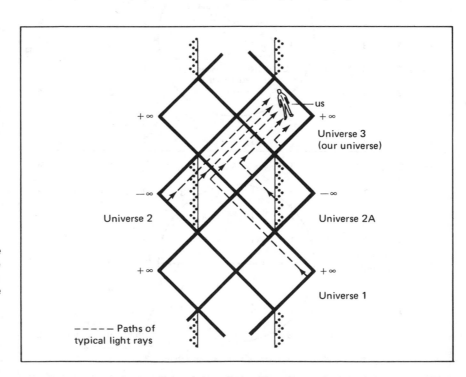

FIGURE 12-18. *The Penrose Diagram for a Kerr Black Hole with $M > a$. When a black hole rotates at a rate less than the extreme speed, light from the singularity bounding a second negative universe (Universe 2A) is reflected inside the hole to a distant astronomer in our universe.*

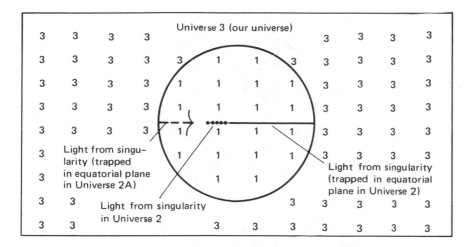

				Universe 3 (our universe)							
3	3	3	3					3	3	3	3
3	3	3	3	3	1	1	3	3	3	3	3
3	3	3	3	1	1	1	1	3	3	3	3
3	3	3	3	1	1	1	1	3	3	3	3
3	3	3	3	1	1	1	1	3	3	3	3
3				1	1	1	1	3	3	3	3
3					1	1		3	3	3	3
3	3						3	3	3	3	
3	3				3	3	3	3	3	3	

Light from singu-
larity (trapped
in equatorial plane
in Universe 2A)

Light from singularity
in Universe 2

Light from singularity
(trapped in equatorial
plane in Universe 2)

FIGURE 12-19. *The Appearance of a Nonextreme Kerr Black Hole at $\theta = 90°$.* Looking in along the equatorial plane at a nonextreme ($M > a$) rotating black hole, an astronomer can see light from a second negative universe (Universe 2A) reflected toward him from inside the hole.

still receives reflected light from Universe 1 (the previous positive-space universe) and from the distant past in his own universe. However, since the hole is rotating slower, another negative-space universe appears in the Penrose diagram. Light from the singularity bounding this additional negative universe (Universe 2A) is also reflected inside the hole out to the distant astronomer. He therefore can see light from the singularity of Universe 2A. These light rays reach the distant astronomer only if the hole is viewed along the equatorial plane ($\theta = 90°$). Figure 12-19, based on calculations by C. T. Cunningham, shows the appearance of a near-extreme Kerr black hole ($a = 90\% M$). The view is essentially the same as in the extreme case (see Figure 12-17C) except that light from the singularity bounding the second negative universe is seen. Light from this second singularity appears slightly to the left of the field of view and includes two small "cusps" that extend slightly above and below the equatorial plane.

A final and fascinating set of exercises deals with what adventurous astronauts would see as they plunge in and out of rotating Kerr holes. Consider, first of all, a *suicide trip.* Two astronauts leave our universe and plunge into a non-extreme Kerr black hole along the equatorial plane. Since they aim their spacecraft along the equatorial plane, they realize they will hit the singularity and be torn apart by infinitely curved space-time. They decide to continue anyway.

Figure 12-20 shows the path followed by the suicidal astronauts. The astronauts travel straight in to the singularity along the equatorial plane. Following the notation used in earlier discussions, the trip

207

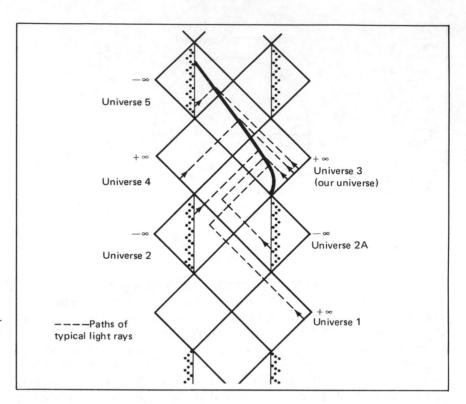

FIGURE 12-20. *The Suicide Trip.* The path followed by two suicidal astronauts is shown in the figure on a Penrose diagram. The astronauts pilot their spacecraft along the equatorial plane of a nonextreme Kerr black hole ($M > a$).

originates in our universe (Universe 3). As with the astronomer who views the black hole, the astronauts can see light from Universe 2, Universe 2A, and Universe 1. In addition, when they cross the outer event horizon, they will be able to see Universe 4, a positive-space universe that is opposite our universe in the Penrose diagram. Finally, upon crossing the inner event horizon, they will see Universe 5, a negative-space universe bounded by the singularity which they will not survive. These various universes and paths followed by typical light rays are shown in Figure 12-20.

In order to take the trip, the astronauts build a spacecraft. The spacecraft has two large windows, as shown in Figure 12-21. Exactly one-half the entire sky can be seen from each window. In designing the spacecraft, the astronauts realize that they have a potential problem. Once inside the ergosphere, the dragging of inertial frames will cause the spacecraft to spin violently. To avoid this difficulty, they equip their vehicle with stabilizing rockets. These rockets ensure that the front window is always pointed toward the singularity, and the rear window pointed toward the outside universe.

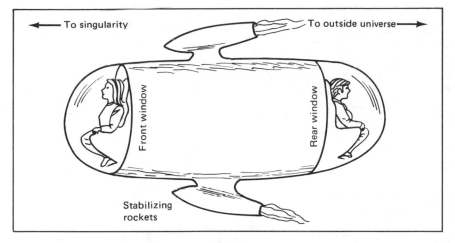

FIGURE 12-21. *The Suicide Spaceship.* Exactly one-half the sky can be seen from each of two windows on the spaceship. The spaceship is equipped with stabilizing rockets which ensure that the front window is always pointed toward the singularity and the rear window always faces in exactly the opposite direction.

During the course of their trip, the astronauts take pairs of photographs, one showing the front view and one showing the rear view, at various significant points along their descent. Figure 12-22 shows the locations of seven such pairs of photographs along their path in a Penrose diagram.

The astronauts jump into their spaceship and blast off. When they are still very far from the black hole, they take the first pair of photographs (Figure 12-23, view A). Since they are still in nearly flat space-time, the front view looks very similar to the view of an astronomer safely back on earth (compare this view A and Figure 12-19). The scene from the rear window is not terribly interesting. From

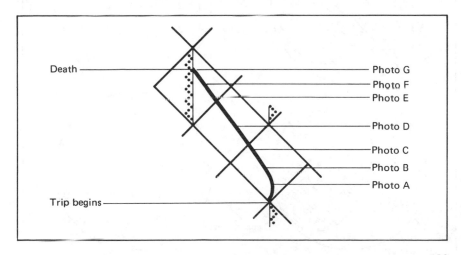

FIGURE 12-22. *The Locations of Photographs.* Pairs of photographs (one for the front view, another for the rear view) are taken by the astronauts at seven locations during their descent.

209

FIGURE 12-23A through G.
A: *Far from the Black Hole.* Far from the black hole, the front view is similar to that seen by a distant astronomer. The rear view simply shows our universe, from which the astronauts came.

B: *Inside Circular Light Orbits.* Once inside the circular light orbits, the astronauts see light rays trapped by the black hole's gravitational field. From the rear window, they still see our universe although images are now somewhat distorted.

C: *Just above the Outer Event Horizon.* As the astronauts approach the outer event horizon, the region occupied by trapped light grows while the region occupied by Universe 1 shrinks. Although only our universe can be seen from the rear window, images are now severely distorted.

D: *Between the Inner and Outer Event Horizons.* Upon passing through the outer event horizon, the astronauts see light from Universe 4. In addition, some of the trapped light is pulled around to the rear view.

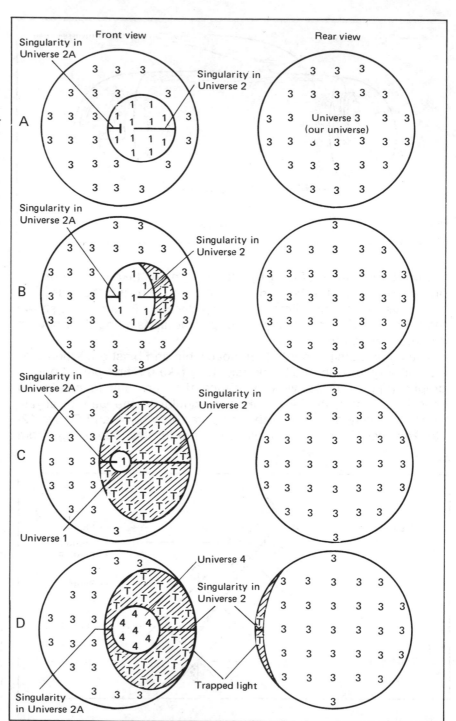

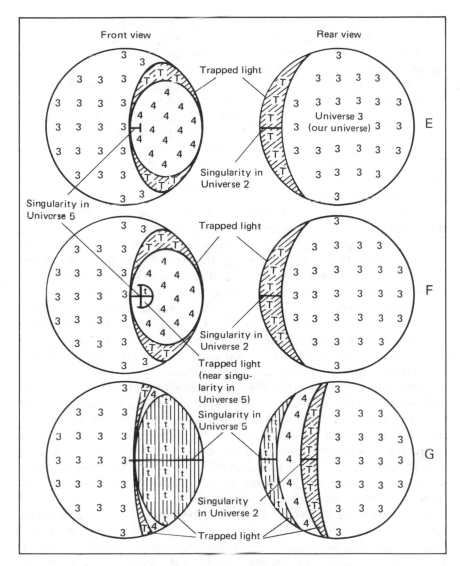

Front view   Rear view

Trapped light

Universe 3 (our universe)

Singularity in Universe 2

Singularity in Universe 5

E

Trapped light

Singularity in Universe 2

F

Trapped light (near singularity in Universe 5)

Singularity in Universe 5

Singularity in Universe 2

Trapped light

G

FIGURE 12-23 *continued.*
E: *Just inside the Inner Event Horizon.* Once the astronauts are inside the inner event horizon, the singularity in Universe 5 comes into view. A large region occupied by reflected light from Universe 4 is seen, and more trapped light is pulled around into the rear view.

F: *Between the Inner Event Horizon and the Singularity.* Inside the inner event horizon, a new region of light trapped near the singularity is seen. Only on a suicide mission can this second region of trapped light be seen.

G: *Just above the Singularity.* Just before the astronauts hit the singularity, the new region of trapped light has grown so large that it is pulled around into the rear view. The view shown here is the last thing the astronauts see.

the rear window, the astronauts simply see where they came from.

As the astronauts approach the black hole, they enter the region of circular light orbits. Inside this region they see light trapped by the hole's gravitational field. In the second pair of photographs (view B, Figure 12-23), the portion of the field of view occupied by the trapped light is indicated by T's. From the rear window only our universe is seen, but specific images of stars and galaxies are now somewhat distorted.

211

As the astronauts fall closer to the outer event horizon, the portion of the field of view occupied by the trapped light gets bigger and bigger. In addition, the astronauts are getting farther and farther away from Universe 1. The portion of the field of view occupied by Universe 1 gets smaller and smaller, as seen in the pair of photographs taken just above the outer event horizon (view C, Figure 12-23). In fact, the moment the astronauts pass through the outer event horizon, Universe 1 completely disappears from sight. All the way up to the outer event horizon, only Universe 3 has been visible from the rear window. But as the astronauts get closer and closer to the black hole, the images of stars and galaxies in our universe are more and more highly distorted.

To understand what happens after the astronauts pass through the outer event horizon, refer back to the Penrose diagram in Figure 12-20. Recalling that light rays travel along 45° lines, we can see that no light from Universe 1 reaches the spaceship once it passes through the outer event horizon. But also notice that between the inner and outer event horizons, the astronauts can see light rays directly from Universe 4. Therefore, photographs taken in between the two horizons (view D, Figure 12-23) contain light from stars and galaxies in Universe 4. In addition, the dragging of inertial frames is now so severe that some of the trapped light is pulled around to the rear view.

After falling through the inner event horizon, the astronauts can still see light from Universe 4. This light is reflected from inside the hole in exactly the same way that light from Universe 1 was reflected. Just inside the inner event horizon (view E, Figure 12-23), the singularity bounding Universe 5 can be seen. This is the singularity that the astronauts are destined to hit. Also notice that the singularity bounding Universe 2A is no longer visible while more trapped light is pulled around into the rear view.

As the astronauts proceed further, the singularity in Universe 5 appears to grow in size. Also, they notice that the singularity is surrounded by another region of trapped light, indicated by t in their next pair of photographs (view F in Figure 12-23). In addition, the light from Universe 3 is now coming to the astronauts in two ways. They see light directly from our universe, primarily through the rear window, but also receive light from our universe which bounces off the singularity in Universe 5.

Finally, just before the astronauts are torn apart at the singularity, this second region of trapped light has grown so large that it is pulled around into the rear view. The final pair of photographs (view G, Figure 12-23) is the very last thing the astronauts see just before they are destroyed. Also notice that most of the reflected light from Universe 4 has been pulled around to the rear view.

Upon hearing about the suicide mission of their colleagues, two additional astronauts remain fascinated by the prospects of space travel through a black hole but devise a safer trip. Since they understand the ring nature of the Kerr singularity, the astronauts decide to fall into the black hole along the axis of rotation. In this way they will escape the devastating effects of infinitely warped space-time. Furthermore, the astronauts decide that their spacecraft shall have *no* rockets. Once the plunge begins, they will *fall freely* into the rotating black hole. The fact that the astronauts are falling freely along the axis of rotation has an important implication. When they arrive at the ring singularity, they will experience antigravity so strong that they will *bounce* back out of the hole. Only if they had rockets blasting could they force their way into the negative universe on the other side of the ring. This mission is therefore called the *bounce trip*.

Figure 12-24 shows the bounce trip on a Penrose diagram. Using the notation previously adopted, the mission begins in Universe 3 (our universe). In plunging through the hole, the astronauts bounce off the singularity in Universe 5. At the end of the trip, they emerge into Universe 7.

The design of the spacecraft used in this mission is shown in Figure 12-25. Notice that the vehicle has no rockets. By allowing the spacecraft to fall freely into the hole, the astronauts are assured that they will bounce out again due to the antigravity they experience at the ring singularity. The spacecraft has two large windows from each of which exactly one-half the entire sky can be seen. The astronauts decide that the window that *always* faces the singularity (both before *and* after the bounce) shall be called the front window. Conversely, the window that always faces the outside universe (toward Universe 3 before the bounce and toward Universe 7 after the bounce) shall be called the rear window.

During the course of the trip, the astronauts take eleven pairs of photographs (views A through K) showing significant changes in the appearance of the sky. As in the case of the suicide mission, the views seen by the astronauts are based on calculations by Cunningham. Figure 12-26 shows the locations of the photographs at various points of the trip on a Penrose diagram.

Confident they will survive the mission, the two astronauts jump into their spacecraft and begin the descent. While still far from the black hole, they take the first pair of photographs (view A, Figure

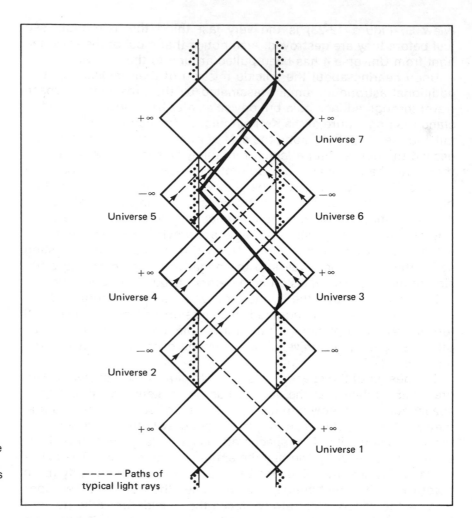

FIGURE 12-24. *The Bounce Trip.* The path followed by two astronauts is shown here on a Penrose diagram. Their spaceship falls along the axis of rotation of a nonextreme Kerr black hole (*M>a*).

Within the diagram:

+∞ — Universe 7
−∞ — Universe 6
Universe 5 — −∞
+∞ — Universe 3
Universe 4 — +∞
−∞ — Universe 2
+∞ — Universe 1

- - - - - Paths of typical light rays

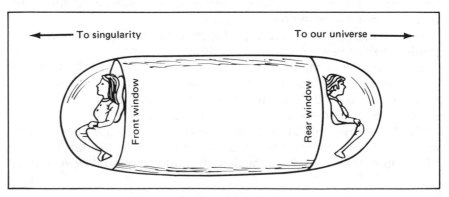

FIGURE 12-25. *The Spaceship.* The spaceship used by the astronauts traveling along the axis of rotation does *not* have any rockets. There are two windows from each of which exactly one-half the sky can be seen. The front window is always pointed toward the singularity—both before *and* after the bounce.

◄──── To singularity   To our universe ────►

Front window   Rear window

214

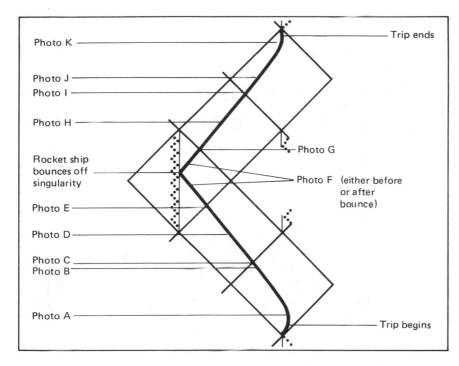

Photo K ——————

Photo J ——————
Photo I ——————

Photo H ——————

Rocket ship
bounces off ——————
singularity

Photo E ——————

Photo D ——————

Photo C ——————
Photo B ——————

Photo A ——————

— Trip ends

— Photo G

— Photo F (either before
or after
bounce)

— Trip begins

FIGURE 12-26. *The Locations of Photographs.* Pairs of photographs (one for the front view toward the singularity, and one for the rear view toward the outside universe) are taken by the astronauts at eleven locations along their journey.

12-27). Since they are still in nearly flat space-time, their view is very similar to that seen by a distant astronomer. (Compare Figures 12-27A and 12-17A.)

As the astronauts approach the outer event horizon, they are getting farther and farther from Universes 1 and 2, which get smaller and smaller. Indeed, both of these universes disappear forever once the outer event horizon is crossed. This fact can be verified by examining the paths of typical light rays in the Penrose diagram shown in Figure 12-24. Just before crossing the outer event horizon, the astronauts notice that they can see trapped light circling the black hole, as revealed by their second pair of photographs (view B, in Figure 12-27).

As mentioned above, Universes 1 and 2 disappear when the astronauts cross the outer event horizon. As seen in the third pair of photographs (view C, Figure 12-27), the central part of the front view is filled only with trapped light.

When the astronauts move inside the outer event horizon, they can now receive light from Universe 4. This fact can be easily confirmed by examining the paths of typical light rays in the Penrose diagram shown in Figure 12-24. A pair of photographs taken be-

215

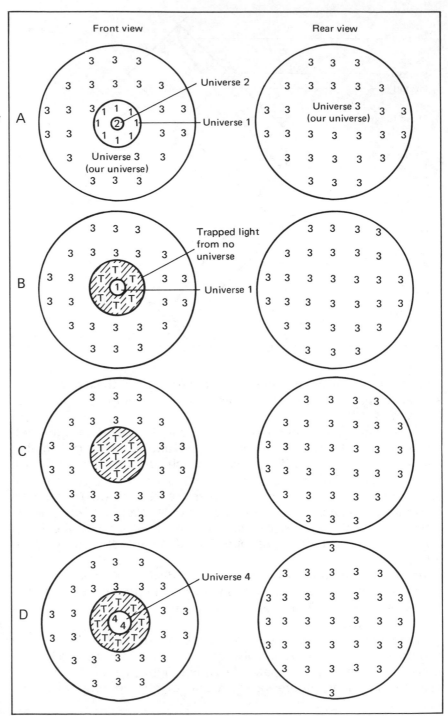

FIGURE 12-27A through K.
A: *Far from the Black Hole.* Far from the black hole, the front view is very similar to that seen by a distant astronomer. The rear view simply shows our universe, from which the astronauts came.

B: *Just above the Outer Event Horizon.* As the astronauts approach the outer event horizon, the apparent sizes of Universes 1 and 2 rapidly shrink. The astronauts also see a region of trapped light from their front window. The view of our universe from the rear window has become distorted.

C: *At the Outer Event Horizon.* At the outer event horizon, Universes 1 and 2 disappear completely. The central part of the front view is filled only with trapped light, while the scene of our universe from the rear window is even more distorted.

D: *Between the Inner and Outer Event Horizons.* Between the two horizons, light from Universe 4 can be seen at the center of the view from the front window. Distortions in the images from our universe (Universe 3) continue to increase.

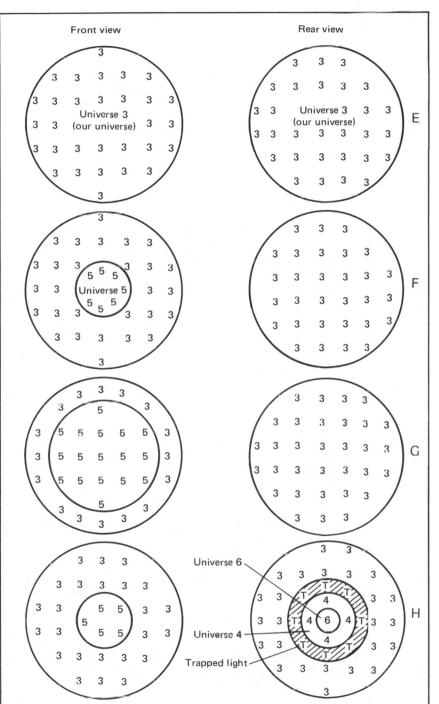

Front view | Rear view

FIGURE 12-27 *continued.*
E: *At the Inner Event Horizon.* At the inner event horizon, only light from our universe (Universe 3) can reach the astronauts. The images of stars and galaxies are, however, very severely distorted.

F: *Near the Time of the Bounce.* Inside the inner event horizon, light from Universe 5 can be seen through the window facing the singularity. Light from the past, present, and future of our universe (Universe 3) is all mixed up.

G: *At the Inner Event Horizon.* As the astronauts cross the inner event horizon on the journey back out of the black hole, light from Universe 5 fills most of the field of view from the front window, which faces the singularity.

H: *Between the Inner and Outer Event Horizons.* As the astronauts recede from the black hole after the bounce, Universes 6 and 4 come into view through the outward-facing (rear) window. The region of light from Universe 5, seen through the inward-facing (front) window, gets smaller and smaller.

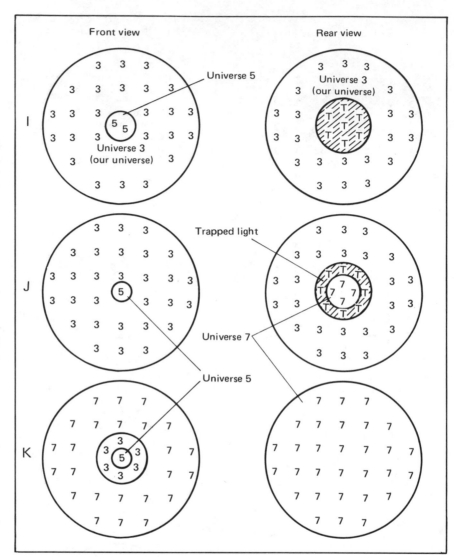

FIGURE 12-27 *continued.*
I: *At the Outer Event Horizon.*
No light from Universe 4 or 6 can reach the astronauts at or beyond the outer event horizon. The image of Universe 5 from which they are receding continues to decrease in size.

J: *Just above the Outer Event Horizon.* Upon emerging through the outer event horizon, the astronauts can see Universe 7 at the center of the outward-facing (rear) window. Universe 5 gets still smaller.

K: *Far from the Black Hole.* Very far from the black hole, Universe 7 fills the entire view through the outward-facing (rear) window. The view toward the black hole is the same as that seen by an alien astronomer living on a planet in Universe 7.

tween the two horizons (view D, Figure 12-27) therefore shows a region at the center of the front view occupied by light from Universe 4.

As the astronauts pass from the outer to the inner event horizons, the region occupied by light from Universe 4 initially increases but then begins to decrease. Indeed, at the inner event horizon, Universe 4 disappears completely. The entire field of view from both front and rear windows is filled only with light from Universe 3. Light from no other universes can reach the astronauts as they cross the inner event horizon, as revealed by the pair of photographs taken at this location (view E, Figure 12-27). By now, however, the images of stars and galaxies in our universe have become severely distorted.

Upon crossing through the inner event horizon, the astronauts can now see light from Universe 5. Both before and after the bounce, Universe 5 is seen surrounding the center of the field of view from the front window. It is important to remember that both before and after the bounce, the front window *always* faces the singularity. Conversely, the rear window *always* faces the outside universe (Universe 3 on the way in and Universe 7 on the way out). Therefore the pair of photographs taken near the time of the bounce (view F, Figure 12-27) apply to either just before or just after the bounce.

As the astronauts recede from the singularity on their way back out of the hole, the region occupied by light from Universe 5 continues to grow in size. As revealed in the pair of photographs taken at the inner event horizon (view G, Figure 12-27), Universe 5 fills almost the entire field of view of the front window, which faces the singularity.

As the astronauts pass outward through the inner event horizon, Universe 6 now comes into view. In addition, they can also see reflected light from Universe 4. This light is reflected off of the singularity that bounds Universe 6, as verified by examining the paths of typical light rays in the Penrose diagram in Figure 12-24. The pair of photographs taken in between the two horizons (view H, Figure 12-27) also demonstrate that the astronauts again see trapped light. And, finally, since they are getting farther and farther from Universe 5, its image is now decreasing in size.

From examining the paths of typical light rays in the Penrose diagram in Figure 12-24, it is seen that no light from Universe 4 or 6 can reach the astronauts at the outer event horizon. Therefore, when they reach the outer event horizon, these two universes have vanished from view. As shown in the pair of photographs taken at this location (view I, Figure 12-27), trapped light fills the central region of the outward-facing (rear) window. Also notice that Universe 5, from which they are receding, appears still smaller.

After emerging through the outer event horizon, the astronauts are in Universe 7, where their trip will end. Universe 7 therefore begins to appear at the center of the outward-facing (rear) window, as shown in the pair of photographs taken just above the outer event horizon (view J, Figure 12-27). As the astronauts get farther and farther from the hole, more and more of Universe 7 is seen. In addition, the image of Universe 5 continues to decrease in size, as does the image of our universe (Universe 3). Finally, therefore, very far from the black hole, Universe 7 completely fills the outward-facing (rear) view. The light from Universes 3 and 5 occupies only a tiny portion of the center of the inward-facing (front) window. The pair of photographs taken at the end of the trip (view K, Figure 12-27) are the same as the view seen by an alien astronomer living in Universe 7.

In conclusion, it should be noted that all the pairs of photographs taken by either the suicidal or bounced astronauts do not really show the whole story. To avoid hopelessly confusing the diagrams, no attempt has been made to display the numerous and complicated image distortions of stars and galaxies that the astronauts really see. In addition, no attempt has been made to display the numerous and complicated redshifts and blueshifts of light which the astronauts also observe. Nevertheless, it will be seen in a later chapter that the existence of highly blueshifted light at the event horizons has some important and profound implications.

13 OBSERVATIONS OF BLACK HOLES

At first glance, the possibility of actually discovering black holes in space seems hopeless. Nothing, not even light can escape from a black hole. It is therefore absurd to suppose that astronomers sitting at their telescopes might ever see a black hole in the sky.

Nothing can ever get out of a black hole because it possesses an intense gravitational field. However, through its gravitational field, a black hole could exert considerable influence on the behavior of nearby objects. Perhaps, therefore, a black hole might be discovered by noticing unusual behavior of visible objects moving in its vicinity. This would amount to an *indirect* discovery of a black hole and great care should be taken to rule out non-black-hole interpretations of the data.

A large percentage of the stars astronomers see in the sky are really *double stars.* As mentioned in Chapter 6, a double star actually consists of two stars revolving about their common center just as the earth and the moon orbit each other.

There are several ways in which astronomers detect the existence of double stars. First of all, observations extending over many years might reveal two visible stars slowly moving about each other in the sky. This is called a *visual binary.* On the other hand, the astronomer might see only one of the two stars in a binary. Often the second star is simply too faint to be detected. Upon examining spectra of what seems to be a single star, the astronomer notices that the spectral lines shift back and forth with great regularity. This is again conclusive evidence for a binary system, even though only one star is seen. As the visible star approaches the earth, the spectral lines are slightly blueshifted. Half an orbit later, as the visible star recedes from the earth, the spectral lines are slightly redshifted. This is called a *spectroscopic binary.*

Sometimes a binary star is oriented in space so that one star passes directly in front of the other. Owing to this fortuitous orientation of the binary's orbit, the two stars alternately eclipse each other, as seen from the earth. Even though it may be impossible to see the individual stars, at the moment of eclipse the nearest star is blocking some of the light from the more distant star. The earth-based astronomer therefore observes a temporary decrease in the total brightness at the time of each eclipse. This is called an *eclipsing binary.*

Binary stars are important to the astronomer because it is often possible to calculate (or at least estimate) the masses of the stars from detailed observations. As mentioned in Chapter 6, analyses of binary star observations led to the *mass-luminosity relation* (see Figure 6-5), which shows that there is a direct correlation between the masses and brightnesses of main-sequence stars.

In the early 1960s, an interesting idea occurred to two Soviet astronomers, Zel'dovich and Guseynov. Suppose one star in a binary system is a black hole. Such a system would not be a visual binary because, obviously, the black hole is invisible. It would also probably not be detected as an eclipsing binary. Even if the orbit happened to be auspiciously oriented, the black hole would be too small to obscure a significant fraction of the visible star's light. But it could be a spectroscopic binary. As the visible star orbits the black hole, the spectral lines would be alternatively redshifted and blueshifted. Of course, the secondary star in a spectroscopic binary may seem invisible simply because it is very dim. To cope with this non-black-hole interpretation of the data, recall that massive stars are usually the brightest stars. Therefore, if the analysis of the observations of a spectroscopic binary reveals that the visible star is the less massive, the unseen (more massive) companion might be a black hole.

Zel'dovich and Guseynov examined observations of spectroscopic binaries dating back many years. Unfortunately they could not honestly come to any firm conclusions concerning the possible existence of black holes in binary stars.

In the late 1960s, Trimble and Thorne at Caltech tried again. Lists of spectroscopic binaries were scrutinized and they came up with eight possible candidates. In all cases it seemed as though the unseen companion was unusually massive. But once again, Trimble was able to argue that the invisible star in each case did not have to be a black hole. It was possible to imagine that the massive unseen star is really an ordinary star that just happens to be exceptionally dim. In view of the feasibility of non-black-hole interpretations of the data, no firm conclusions could be reached. Finding black holes in binary stars looked impossible.

On Saturday, December 12, 1970, a satellite was launched into earth orbit from a platform in the ocean off the coast of Kenya. This date was the seventh anniversary of Kenyan independence and in appreciation of the hospitality of the Kenyan people, the satellite (Explorer 42) was christened *Uhuru,* which means "freedom" in Swahili. Unlike any other satellite ever launched, Uhuru is entirely devoted to x-ray astronomy (see Figures 13-1 and 13-2). Two x-ray telescopes scan the skies and send back signals to earth everytime an x-ray source is seen. Prior to the launch of Uhuru, astronomers made x-ray observations from small rockets. During the precious moments that the rocket was above the obscuring earth's atmosphere, a few x-ray stars could be observed before the instrument payload plummeted back to earth. From Uhuru, however, continuous observations could be made hour after hour and day after day.

By 1974, a total of 161 x-ray sources had been examined and cataloged. Some of these sources are associated with distant galaxies far beyond our Milky Way, while others appear to be relatively nearby at typical stellar distances. Specifically, at least 8 of the 161 sources are in binary star systems. They are listed in Table 13-1. In the table, 3U stands for the Third Uhuru Catalogue; the remaining numbers denote an approximate position in the sky. The common names often come from the pre-Uhuru days when observations were made from brief rocket flights.

223

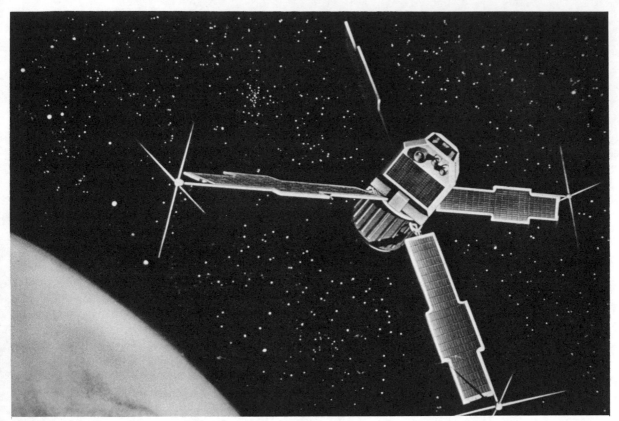

FIGURE 13-1. *Uhuru.* This satellite detects x-rays from stars and galaxies. Since its launch in December of 1970, Uhuru has discovered almost 200 x-ray sources in the sky. (NASA)

FIGURE 13-2. *Uhuru's Instrumentation.* The satellite is equipped with two x-ray telescopes and receives energy from four solar panels. As the satellite slowly spins about its axis, the telescopes scan the skies for x-ray sources.

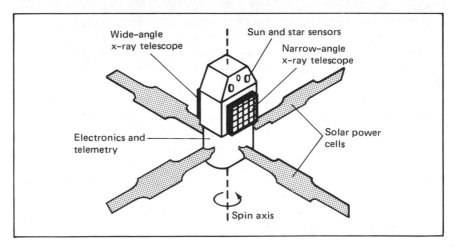

Wide-angle x-ray telescope

Sun and star sensors

Narrow-angle x-ray telescope

Solar power cells

Electronics and telemetry

Spin axis

224

TABLE 13-1 X-ray Stars in Binary Systems

Name	Common Name	Distance (light-years)	Orbital Period (days)
3U 0115 − 73	SMC X-1	190,000	3.9
3U 0900 − 40	Vela X-1	8,000	8.9
3U 1118 − 60	Centaurus X-3	25,000	2.1
3U 1617 − 15	Scorpius X-1	?	0.8
3U 1653 + 35	Hercules X-1	16,000	1.7
3U 1700 − 37	(none)	9,000	3.4
3U 1956 + 35	Cygnus X-1	10.000	5.6
3U 2030 + 40	Cygnus X-3	30,000	4.8

Since all of these eight sources are members of binary systems, astronomers in the early 1970s were faced with the fact that stars can be powerful sources of x-rays. But ordinary stars seen in the nighttime sky do not emit x-rays. Thus there must be something unusual about these x-ray stars. Each of these stars emits about ten thousand times more energy in x-rays than the sun emits at all other wavelengths combined.

One important clue to the nature of some of these x-ray stars came with the discovery of x-ray pulsations from four of the sources. Four of these stars are *x-ray pulsars* and therefore presumably are rotating neutron stars. They are SMC X-1 (pulse period 0.716 second), Vela X-1 (pulse period 282.9 seconds), Centaurus X-3 (pulse period 4.842 seconds) and Hercules X-1 (pulse period 1.238 seconds). In order to understand the details of the mechanism that produces the x-rays, astrophysicists had to begin with some familiar basic concepts.

Back in the nineteenth century, astronomers spent a lot of time discovering and examining binary stars. While the astronomers were looking up into the skies, mathematicians and physicists were performing calculations to aid in the understanding of binary star systems. In particular, it was discovered that an imaginary figure-eight could be drawn about the two stars in a binary, as shown in Figure 13-3. This figure-eight is important because it shows the domain of the gravitational influence of each star. Specifically, any gases inside one loop of the figure-eight belong to the star in that loop and

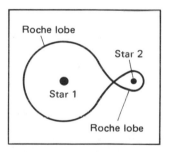

FIGURE 13-3. *The Roche Lobes.* A figure-eight can be drawn about the two stars in a binary. Any gases inside one lobe of the figure-eight belong to the star in that lobe. If, for some reason, gas is ejected out of the Roche lobe, the matter escapes from the star. In this particular drawing, star 1 is more massive than star 2.

225

cannot escape to the other star or off into space. The two loops of the figure-eight are called *Roche lobes* in honor of the person who discovered this useful scheme. If, however, gas is ejected out of a Roche lobe, this matter is then free to wander off into interstellar space. Alternatively, if gas is ejected through the crossover point of the figure-eight, this matter is free to leave one star and fall upon the other star. The crossover point is called the *inner Lagrangian point* and affords a means of *mass transfer* from one star to the other.

Suppose that one star in a binary system is ejecting matter outward through its Roche lobe. This can happen for one of two reasons. During the course of their life cycles, stars sometimes expand to many times their original size. In Chapter 6, it was noted that this effect occurs as stars evolve into red giants. If, as a result of this expansion, the star in a binary system gets bigger than its Roche lobe, the star dumps matter out into space. In this way, one star in a binary can lose a great deal of mass.

The second, more gradual method of mass loss involves *stellar wind.* Astronomers have reason to believe that all stars are constantly ejecting streams of atomic particles into space. For example, using satellites launched from earth, astronomers have detected streams of particles coming from the sun. This is called the *solar wind.* Although solar or stellar winds do not contain very much matter, this gradual leaking of particles into space can have some important consequences. Figure 13-4 illustrates these two methods of mass loss.

As a star in a binary ejects matter through its Roche lobe, some of this material will pass across the inner Lagrangian point and fall toward the second star. If this second star is fairly large, the infalling matter will pile up or *accrete* directly on the star's surface. However, if the second star is very small, the infalling matter will go into orbit about the tiny star. The matter will pile up in a disk or ring about the star, similar to the rings about Saturn. This ring of matter is called an *accretion disk.* Just as Mercury goes around the sun faster than Pluto, the inner edge of an accretion disk is rotating at a faster rate than the outer edge. The fact that different parts of the accretion disk rotate at different speeds means that the gases in the disk are constantly rubbing against each other. This friction both heats these gases and causes them to spiral inward toward the star. If this star happens to be a neutron star with an intense magnetic field, the infalling matter is "funneled" down the north and south magnetic poles. Calculations reveal that this infalling matter strikes the neu-

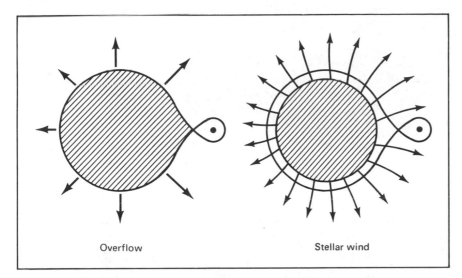

Overflow   Stellar wind

FIGURE 13-4. *Mass Loss*. A star in a binary can lose matter in one of two ways. If the star overflows its Roche lobe by simply expanding to a very large size, great quantities of matter can be dumped into space. Alternatively, particles streaming out from the star's surface in a stellar wind can escape through the Roche lobe. This second method of mass loss is far more gradual than the first.

tron star so violently that vast quantities of x-rays are emitted. This is the essence of the model by which astrophysicists explain x-ray pulsars like Centaurus X-3 and Hercules X-1.

If pulsars can exist in x-ray binaries, what about black holes? What would happen if a black hole rather than a neutron star were at the center of an accretion disk? To examine these fascinating questions, detailed theoretical studies were begun in 1971. Shakura and Sunyaev at Moscow and Pringle and Rees at Cambridge started by using Newtonian theory. Although the calculations were nonrelativistic, it was soon apparent that matter in an accretion disk about a black hole could emit vast quantities of x-rays. This remarkable discovery inspired many other physicists to repeat the calculations more carefully using general relativity. By the mid-1970s, Thorne, Page, and Price at Caltech had succeeded in working out many of the details. The model they obtained is shown in Figure 13-5.

If a binary system contains a black hole and if the normal star is dumping matter over its Roche lobe, an accretion disk will form around the black hole. As gases pass across the inner Lagrangian point, they are captured into orbit about the black hole. It is expected that the disk would have a diameter of several million miles, but would be less than 100,000 miles thick. The flattened appearance of the disk is caused by centrifugal and gravitational forces acting on the gases. Since the calculations concerning the gravita-

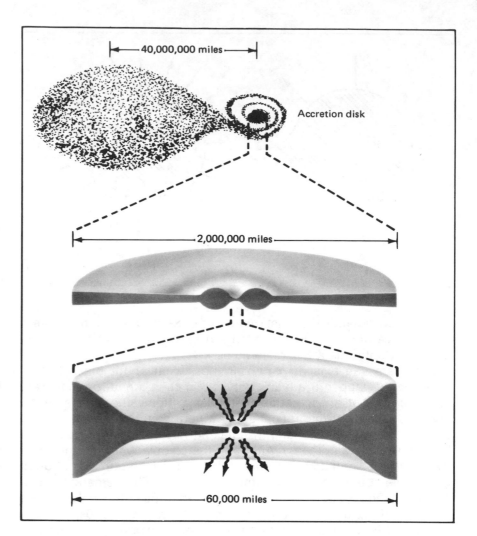

FIGURE 13-5. *An Accretion Disk about a Black Hole.* According to calculations by Thorne and others, an accretion disk about a black hole in a binary system should be about 2 million miles in diameter. X-rays are emitted from the innermost parts of the disk, at a distance of only 200 miles from the hole itself. The dimensions given in this illustration are those expected in the case of the Cygnus X-1 system.

tional field of black holes in accretion disks are to be as realistic as possible, it is entirely reasonable to suppose that the hole is rotating and therefore is described by the Kerr solution.

When gases are initially captured at the outer edge of the accretion disk, their temperature is roughly the same as the surface temperature of the normal star from which they came. Friction between filaments of gas orbiting the hole at various distances soon begins to heat up the infalling gas. As a result of this friction, the temperature rises as the gases spiral inward. As the temperature of the in-

228

ward-spiraling gases increases, the pressures inside the disk also go up. This increasing gas pressure tries to make the disk thicker. However, throughout most of the disk, the gravitational field of the rotating black hole is sufficiently strong to keep the disk quite thin. Only at a distance of roughly 50,000 miles from the hole are the gas pressures powerful enough to thicken the disk. The inner portion of the accretion disk therefore has a "bulge" about 100,000 miles in diameter.

Typically, it would take several weeks or months for some gas to spiral all the way in from the outer edge to the inner edge of the accretion disk. By the time the gas is only a few hundred miles above the black hole, friction has heated the gases to temperatures of ten million degrees. Anything with temperatures this high naturally emits great quantities of x-rays. In fact, x-rays are produced in such profusion that any nearby black hole surrounded by an accretion disk should be detected by Uhuru. At the inner edge of the accretion disk, the hole's gravitational field is so intense that the inward-spiraling gases are sucked into the hole in a fraction of a second.

In examining the list of eight x-ray binary stars, the four pulsars can be immediately eliminated as black-hole candidates. Nothing associated with a black hole could produce regular pulsations and the neutron star interpretation of the data is far more reasonable. But what about the other four? They might be black holes. But they also could be neutron stars (oriented so that we on earth do not see the pulsations) or perhaps even white dwarfs at the center of accretion disks. How can the astronomer distinguish conclusively between these various possibilities?

At this point two important facts should be realized. First of all, in Chapter 7 it was noted that there are strict upper limits to the masses of white dwarfs and neutron stars. The Chandrasekhar limit for white dwarfs is 1¼ solar masses and degenerate neutron pressure cannot support a dead star whose mass is greater than three times that of the sun. Any dead star whose mass is greater than three solar masses *must* be a black hole.

Second, from detailed examination of the orbits of binary stars it is often possible to get reliable estimates of the stars involved. In other words, from observing the orbits of the stars in an x-ray binary it should be possible to calculate the masses of the two stars in each case. If the star at the center of the accretion disk has a mass greater than three suns, it must be a black hole.

In principle, these schemes for finding black holes sound easy. In

TABLE 13-2   Visible Stars in X-ray Binaries

Name of X-ray Source	Name of Visible Star	Apparent Magnitude of Visible Star
3U 0115 − 73 (SMC X-1)	Sanduleak 160	13
3U 0900 − 40 (Vela X-1)	HD 77851	8
3U 1118 − 60 (Centaurus X-3)	Krzeminski's star	13
3U 1617 − 15 (Scorpius X-1)	(unnamed)	12
3U 1653 + 35 (Hercules X-1)	HZ Herculis	14
3U 1700 − 37	HD 153919	7
3U 1956 + 35 (Cygnus X-1)	HDE 226868	9

practice, they are extremely difficult. To begin with, astronomers must see *two* stars in each binary. Of course, one star emits only x-rays and therefore its spectral lines will not appear in an optical spectrum taken at the telescope. In addition, astronomers must find the *visible* star associated with each x-ray binary they wish to examine. During the early 1970s, therefore, astronomers worked very hard trying to find the visible stars associated with x-ray binaries. Only then could ordinary earth-based telescopes be used to study visible components in the hope of obtaining estimates of the mass of the x-ray stars. The search for the visible stars began by obtaining accurate positions for each of the eight x-ray sources. Once accurate positions were available, astronomers scanned the sky near each position looking for a visible star that exhibited the tell-tale signs of being a member of a binary system. In seven of the eight cases, astronomers were successful in identifying a visible star that met all the necessary requirements. These seven visible stars are listed in Table 13-2.

By the mid-1970s, detailed observations of each of the seven visible stars in the x-ray binaries had been completed. Since only one of the two stars in each system could be seen with ordinary telescopes, astronomers had to resign themselves to obtaining estimates of the masses of the x-ray stars. Only when both stars are visible can astronomers calculate precise masses. As previously mentioned, four of the seven binaries contain x-ray pulsars, and estimates of the masses of the x-ray stars came out to be about two times that of the sun. This result is in agreement with the idea that each of the x-ray pulsars consists of a neutron star at the center of an accretion disk.

Of the three remaining candidates, in two cases no firm conclusions could be drawn. The masses of the x-ray stars at most turned out to be "a few" times the sun's mass. The masses of the x-ray stars in 3U 1617 − 15 and 3U 1700 − 37 are *not* conclusively above the critical value of three solar masses. Cygnus X-1, however, is a different story.

Cygnus X-1 was shown to be associated with a hot (25,000 °K), bluish star called HDE 226868 (see Figure 13-6). This star is a spectroscopic binary, and its spectral lines shift back and forth with a period of about 5½ days. Stars that are hot and bluish usually have very large masses. The mass of HDE 226868 is probably greater than twenty solar masses. From observing the shifting spectral lines, and from assuming a mass for the visible star of at least twenty suns, reasonable limits on the mass of Cygnus X-1 could be calculated. Cygnus X-1 must be at least an eight-solar-mass object! Since this is unquestionably far above the maximum mass of a neutron star, it seems entirely reasonable to conclude that *Cygnus X-1 must be a black hole!*

Everything known about Cygnus X-1 can be explained and understood in terms of a black hole at the center of an accretion disk. Al-

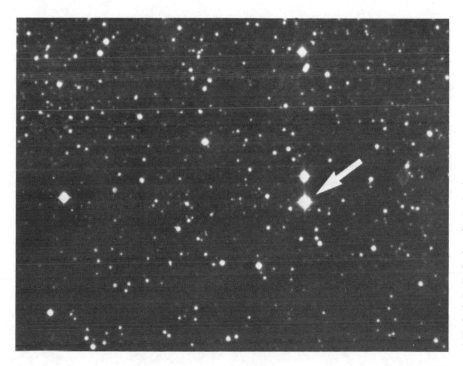

FIGURE 13-6. *HDE 226868.* The visible star associated with Cygnus X-1 is a hot, bluish star at a distance of about 10,000 light-years. From examining this star, astronomers conclude that Cygnus X-1 probably has a mass greater than eight suns. (Copyright by the National Geographic Society–Palomar Observatory Sky Survey. Reproduction by permission from the Hale Observatories.)

though this does not prove that Cygnus X-1 must be a black hole, the overwhelming preponderance of the evidence favors a black-hole interpretation of the data, although some astronomers still have doubts.

It should be emphasized that evidence for Cygnus X-1 being a black hole largely consists of "negative" tests (for example, it can't be a white dwarf; it can't be a neutron star). Positive evidence would be far stronger. What observations could be made of Cygnus X-1 (or a similar system) which would prove it must be a black hole? What phenomena might be unique to a black hole and could not occur with any other type of object?

As noted earlier, the inner edge of an accretion disk surrounding a black hole must be orbiting the hole at a very high speed. Indeed, the inner edge of the disk probably takes only a few hundredths or thousandths of a second to go once around the hole. It is reasonable to suppose that the inner edge of the accretion disk is not perfectly uniform. It might have some "hot spots." Each time a hot spot orbits the hole, astronomers should see a brief flash of intense x-rays in addition to the x-rays normally emitted by the disk. If these extra flashes were observed more frequently than once every hundredth of a second, this could constitute direct evidence for a black hole.

Unfortunately, Uhuru is not capable of detecting very rapid variations in x-rays. It *seems* as though Cygnus X-1 is emitting rapid flashes, but astronomers will not be sure until better x-ray telescopes are placed in earth orbit. In this regard, it should be noted that Circinus X-1 (3U 1516 − 56) also seems to be emitting rapid flashes of x-rays. Although the visible companion to Circinus X-1 has yet to be identified, the similarities of the x-ray data between Cygnus X-1 and Circinus X-1 are very striking. Astronomers may have already discovered a second black hole.

Since it seems quite probable that Cygnus X-1 is a black hole, astrophysicists have begun detailed calculations to understand the evolution of a binary star system that ends up with a black hole. Since the black hole can be detected only from infalling matter ejected by its normal companion, the two stars must be fairly close to each other. If the stars were widely separated (as usually is the case), the black hole could not capture enough matter to emit x-rays. Attention was therefore focused on the life cycles of *close binaries.*

Suppose two stars are born very near each other and form a close binary system, as shown in Stage 1 of Figure 13-7. Both stars initially

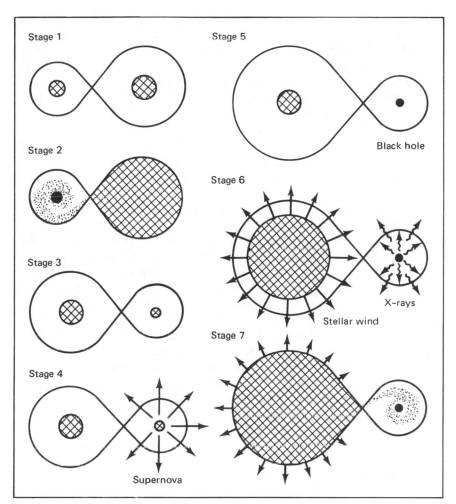

Stage 1

Stage 2

Stage 3

Stage 4

Supernova

Stage 5

Black hole

Stage 6

X-rays

Stellar wind

Stage 7

FIGURE 13-7. *A Black Hole in a Close Binary.* The major stages in the evolution of a close binary system are shown here based on calculations by de Loore and de Greve. The binary emits x-rays only for a brief period during its entire life cycle.

burn hydrogen at their cores, but the more massive star evolves more rapidly. The more massive star soon fills its Roche lobe and dumps great quantities of matter onto its companion, as depicted in Stage 2. Due to this mass transfer, the companion is now the more massive star (Stage 3). After a supernova explosion, a black hole is formed, provided the dying star is left with enough matter to overcome gas pressures that would otherwise form a white dwarf or neutron star (Stage 4 and 5). The black hole remains undetectable until its companion evolves to a stage where it emits a powerful stellar wind. Only then can the black hole capture enough gas to form an

233

accretion disk that emits x-rays, as shown in Stage 6. Finally, when the companion has evolved so far that it fills its Roche lobe, vast quantities of matter are dumped through the inner Lagrangian point onto the black hole. This dumping of matter chokes off the x-rays and the black hole is again undetectable (Stage 7).

This scenario tells astronomers that the x-ray emitting stage of a close binary is very brief. Indeed, close binaries will emit x-rays less than one-half of 1 percent of their entire lifetime. Statistically, therefore, only one out of several hundred close binaries could be expected to be giving off detectable x-rays. For every Cygnus X-1, there easily could be several hundred black holes in close binaries from which no detectable radiation can escape.

In examining data from starlike sources of x-rays while searching for black holes, astronomers announced a remarkable discovery in the spring of 1976 concerning x-ray sources. Using the so-called Astronomical Netherlands Satellite launched into a polar earth orbit on August 30, 1974, astronomers began observing many of the x-ray sources listed in the Third Uhuru Catalogue. On September 28, 1975, while observing the source 3U 1820 − 30 with the new x-ray telescopes on board this satellite, they detected an enormous burst of x-ray radiation. In less than one second the intensity of x-rays from 3U 1820 − 30 increased by a factor of about 25. Over the next eight seconds, the x-ray intensity gradually declined to its preburst level. Several additional bursts from this same source have been observed, a typical recording of which is shown in Figure 13-8. Ener-

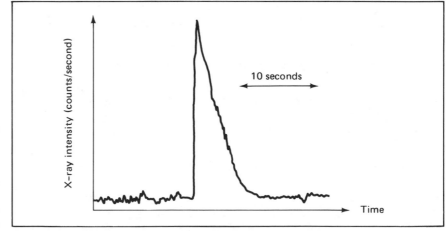

FIGURE 13-8. *An Intense X-ray Burst.* Incredibly intense x-ray bursts from sources in globular clusters have been detected. In less than one second, the intensity of x-rays increases 25-fold. Over the next eight seconds, the intensity drops to its preburst level.

234

FIGURE 13-9. *The Globular Cluster NGC 6624.* Incredibly intense bursts of x-rays have been detected coming from this globular cluster. The most straightforward explanation of these bursts involves a massive black hole at the center of NGC 6624. This photograph by Dr. Neta Bahcall is a short exposure that reveals the stars in the central region of the cluster. (Courtesy of N. Bahcall)

getic bursts of this incredible intensity had not been detected up until this time.

The source 3U 1820 − 30 is associated with the globular cluster NGC 6624. Globular clusters (see Figure 13-9) are huge spherical stellar aggregations typically containing a hundred thousand stars. In trying to explain brief and intense x-ray bursts from a globular cluster, J. Grindlay and H. Gursky at Harvard University concluded that there may be an extremely massive black hole at the center of NGC 6624. Indeed, all of the observations can be easily understood in terms of an accretion disk about a black hole whose mass is roughly 500 suns. Presumably, a sudden change in the accretion disk triggers the x-ray burst, sending an intense flood of radiation outward through the gas cloud, whose diameter is approximately 20 light-seconds.

How such extremely massive black holes might be formed will certainly be a subject of intense theoretical study during the next few years. Globular clusters are among the oldest stellar systems in our galaxy. Perhaps numerous smaller black holes resulting from the deaths of massive stars over the ages coalesce and swallow each other up at the centers of globular clusters. A conglomeration of roughly a hundred smaller black holes would be necessary to account for the incredibly brief and intense x-ray burst from 3U 1820 − 30 in NGC 6624.

14 WHITE HOLES AND PARTICLE CREATION

The possible existence of black holes in space is among the most fascinating predictions of theoretical physics of the twentieth century. The idea that black holes should exist is a direct consequence of our current understanding of stellar evolution. Dying massive stars collapse in upon themselves, producing a region where gravity is so strong that nothing, not even light, can escape.

In examining the theoretical properties of black holes, it was discovered that all black holes must have mass. But in addition, they can have charge and/or angular momentum. In general, a realistic black hole probably has a negligible charge but a high rate of rotation. It would therefore be accurately described by the Kerr solution.

This same theoretical investigation, however, reveals that the full geometry of an ideal black hole is extremely complicated. Indeed, the global structure of space-time connects a variety of uni-

237

verses as revealed in Penrose diagrams. For the simplest black hole containing only matter (a Schwarzschild hole as shown in Figure 9-11 and Figure 9-18) there is one other universe besides our own. Due to the spacelike nature of the Schwarzschild singularity, this other universe cannot be reached along any allowable (timelike) paths from our own universe. But when either charge or rotation is included, the singularity is timelike and the full geometry of the Reissner-Nordstrøm or Kerr solution connects an infinite number of past and future universes (see Figure 10-10 and Figure 11-14). This many-universe nature of the Kerr and Reissner-Nordstrøm solutions affords the incredible possibility of hypothetical space travel in and out of black holes to future universes, as discussed in Chapter 12. It also gives rise to the possibility of a time machine!

There are several different ways of interpreting the "other" universes that appear in a Penrose diagram. One way is to say that they are really separate, distinct universes totally unrelated to our own universe. Another equally acceptable interpretation is that some of these other universes are really different versions of our own universe at different epochs. In other words, it is theoretically possible that one of the other universes in a Penrose diagram is actually *our* universe a billion years ago, as shown in Figure 14-1. A hypothetical astronaut could therefore leave the earth now, plunge into the black hole and come out into our universe at an earlier time. This is time travel.

Similarly, some other universe in the Penrose diagram might actually be our own universe in the very distant future. A hypothetical astronaut could therefore leave the earth and come back billions of years in the future simply by emerging into the appropriate universe on the Penrose diagram.

Although Figure 14-1 is the Penrose diagram for a Kerr black hole (with dotted singularities and negative universes), the same general features apply to a Reissner-Nordstrøm black hole. In either case, by interpreting some of the other universes as different versions of our own universe at different times, we could travel freely into the past and future.

The idea that time machines could exist is extremely troublesome to the scientist. Truly disturbing things could happen. For example, imagine an astronaut who blasts off from earth and plunges into a rotating or charged black hole. After wandering around for a while, he finds a universe which is his universe, but 10 minutes earlier. By coming out into this earlier universe, he finds everything just the way it was a few minutes before he left. Indeed, he could find him-

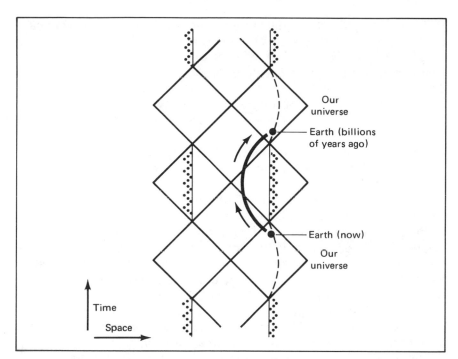

FIGURE 14-1. *A Time Machine.* If one of the other universes in a Penrose diagram is actually our universe at an earlier time, an astronaut could leave the earth now and return to the earth in the past by traveling through a black hole.

Our universe

Earth (billions of years ago)

Earth (now)

Our universe

Time

Space

self just before he boarded his rocketship (see Figure 14-2). He could meet himself and tell himself what a nice trip he had. Then *both* of him could get on the waiting rocketship and he (they?) could both make the same trip again!

FIGURE 14-2. *A Violation of Causality.* By emerging into his own universe at a slightly earlier time, an astronaut could meet himself just before he left. This violates causality.

239

This journey, however, illustrates that time machines violate *causality*. Causality simply states that effects occur *after* their causes. If a light bulb in a room suddenly turns on, it is reasonable to assume that somebody flicked the switch a fraction of a second earlier. It is absurd to suppose that a light bulb could now turn on because somebody 10 years in the future flicks the switch. The idea that effects could occur before their causes is denied by the rational human mind.

There are, therefore, two possibilities. First, perhaps causality can be violated. This would mean that physical reality is irrational at a very fundamental level: The universe is crazy and the appearance of rationality is a complete fiction artificially imposed by the human mind. Perhaps scientists believe in causality in order to understand a universe that can never be understood.

The second possibility is that the Penrose diagrams do not tell the whole story. Perhaps additional physical effects occur that destroy the possibility of traveling to other universes. Perhaps the Penrose diagram is so idealized that it does not describe anything that could really exist in the first place.

Kruskal-Szekeres and Penrose diagrams were invented in order to understand the geometry of space-time of a black hole more fully. From such diagrams, many of the properties of black holes can be appreciated. Yet, the diagrams themselves predict things. For example, Figure 14-3 is the Kruskal-Szekeres diagram for a Schwarzschild black hole. Very simply, stuff from our universe falls in through the event horizon and hits the singularity. However, suppose there were matter and radiation near the past singularity. As time moves on, this matter and radiation will emerge through the past event ho-

FIGURE 14-3. *A Black Hole* (*left*). In the case of a Schwarzschild black hole, all infalling matter and radiation pass through the event horizon and are crushed at the singularity.

FIGURE 14-4. *A White Hole* (*right*). Conceivably, matter and radiation from the region of space-time near the past singularity could emerge into our universe, thereby producing a white hole.

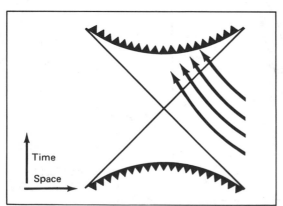

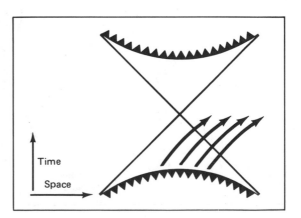

rizon and come gushing into our universe. This is a *white hole,* and is depicted in Figure 14-4.

Finally, imagine matter that gushes up from the region near the past singularity, rises to a certain height above the black hole, and then plunges back in. As shown in Figure 14-5, this behavior of matter is allowed on the Kruskal-Szekeres diagram because the paths are everywhere timelike. An object that behaves in this fashion is called a *grey hole.*

While the concept of black holes came from understanding stellar evolution, the idea of a white hole or grey hole originated out of the mathematics of the Schwarzschild solution. As with time machines, are we to believe that white holes and grey holes really can exist in the universe?

Imagine a massive dying star that is collapsing to form a black hole. Initially there is no singularity. Initially there is no event horizon. Therefore, there cannot be any past singularity or past event horizon. Only the future event horizon and singularity exist because the black hole forms in the future, after the star dies. In other words, as shown in Figure 14-6, the body of the star cuts off a large section of the Kruskal-Szekeres diagram. Only above the surface of the star is space-time properly described by the Schwarzschild solution. Therefore, in the realistic application of the Schwarzschild solution, there cannot be any grey holes or white holes. A collapsing star that becomes a Schwarzschild black hole simply does not have a past singularity or past event horizon. Neither does the "other universe" exist.

Although examining the processes that occur in the death of a star eliminates the possibility of creating Schwarzschild grey holes

FIGURE 14-5. *A Grey Hole* (*left*). Conceivably, matter from the region near the past singularity could emerge into our universe only to plunge back into the hole and strike the future singularity.

FIGURE 14-6. *The Formation of a Black Hole* (*right*). When a dying star collapses to form a Schwarzschild black hole, most of the Kruskal-Szekeres diagram is cut off by the matter of the star.

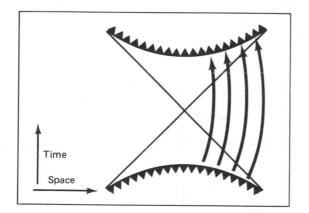

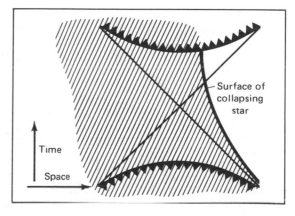

241

and Schwarzschild white holes, our problems are far from over. As we have noted many times, a realistic star would be rotating and therefore form a Kerr black hole. The complete structure of space-time for a Kerr black hole is shown in a Penrose diagram with time-like singularities. If we imagine a real star collapsing to form a Kerr black hole, large sections of space-time are eliminated. The Kerr geometry applies only to those regions of space-time *above* the star's surface. Nevertheless, as shown in Figure 14-7, a star that forms a black hole in one universe could emerge as a white hole in another universe. Due to the timelike nature of the singularity, a collapsing star from one universe could expand into another universe. Thus, the Kerr solution (as well as the Reissner-Nordstrøm solution, which also has timelike singularities) seems to allow for the possible existence of white holes.

The concept of Schwarzschild white holes was revived in the mid-1960s by the Soviet physicist I. D. Novikov. Although Schwarzschild white holes cannot form from dying stars, Novikov began by thinking about the creation of the universe. Most astronomers believe that the universe started with a *big bang,* a primordial explosion from an infinitely dense state. In other words, the entire universe must have been one gigantic singularity which blew up for some unknown reason. Suppose that certain selected regions did *not* participate in the overall expansion of the universe. Suppose that for some reason a little piece of the primordial singularity could survive without expanding for a very long time. When, finally, this "lagging core" did decide to expand, it would have all the properties of a

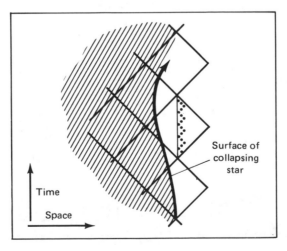

FIGURE 14-7. *A Kerr White Hole.* The formation of a rotating black hole in one universe could create a white hole in another universe.

Surface of collapsing star

Time

Space

242

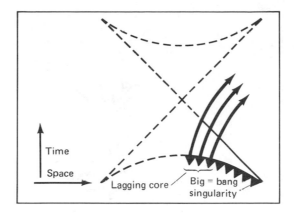

FIGURE 14-8. *A Lagging Core.* If a piece of the big bang did not expand with the rest of the universe, it might erupt later as a Schwarzschild white hole.

white hole. The lagging core is literally a piece of a past singularity (the big bang) out of which matter and radiation could erupt into our universe. The idea that little pieces of the big bang could have survived for a long time led Novikov to propose the possibility that Schwarzschild white holes might exist.

The problem of Schwarzschild white holes was tackled by D. M. Eardley at Caltech in the early 1970s. Dr. Eardley realized that if a lagging core were left over from the big bang, it must look like a piece of a past singularity and therefore be surrounded by a past event horizon (see Figure 14-8). But what do we know about event horizons? In ordinary black holes, an event horizon is where time seems to stop according to a distant observer. To a distant observer, light coming from very near the event horizon appears to have a very large redshift. Crudely speaking, the light from near an event horizon loses a lot of energy in climbing out of the intense gravitational field surrounding an ordinary black hole. Conversely, if light falls into a black hole, the radiation must gain a lot of energy. Infalling light must become very *blueshifted.*

Think for a moment about the very earliest stages of the universe. If there were a big bang, the universe must initially have been extremely hot. At enormous temperatures of trillions of degrees, the entire universe must have been filled with intense radiation. If there were any lagging cores left over from the big bang, this radiation (already very intense) would have become highly blueshifted as it fell on the event horizon surrounding the lagging core. Incredible amounts of highly energetic radiation should have begun piling up around the lagging core. Put another way, in a Penrose diagram light from $\mathcal{I}^-$ piles up along the past event horizon producing a *blue*

243

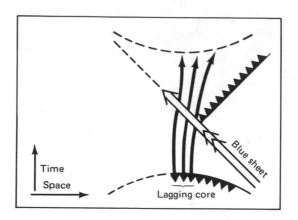

FIGURE 14-9. *The Death of a White Hole.* If a potential white hole could exist as a lagging core left over from the big bang, so much light would be captured that a black hole would be created. The potential white hole would be transformed into a black hole in a very short period of time.

sheet. In a very short period of time, this blue sheet contains so much light that the energy of the light itself begins to warp space-time severely. According to Eardley's calculations, the light piling up around the lagging core warps space-time so effectively that a *black hole forms around the potential white hole.* As shown in Figure 14-9, a future event horizon and singularity form. This transformation of a potential white hole into a black hole typically occurs in about a thousandth of a second. Therefore, if there were any lagging cores, they all would have turned into black holes very shortly after the creation of the universe.

Eardley's calculations effectively dispose of any possibility that Schwarzschild white holes might exist in nature. But what about Reissner-Nordstrøm white holes or Kerr white holes? Although detailed calculations have not yet been made, Eardley's arguments seem highly relevant. In order for one of these more complex white holes to exist, there must be a few inner and outer event horizons through which matter can pass from one universe to the next. In examining a Penrose diagram for a charged or rotating black hole, one quickly realizes that a future event horizon for one universe is at the same time a past event horizon for another universe. The event horizon through which matter falls into a black hole in one universe is the same event horizon out of which matter gushes from a white hole in the next universe. Thus, if Reissner-Nordstrøm or Kerr white holes exist, they must have past event horizons. If a white hole in some universe has a past event horizon, ever since the creation of that universe light has been piling up along the horizon. The horizon *must* have a blue sheet. Following Eardley's arguments, so much light will have accumulated that the energy contained in the blue

244

sheet makes the event horizon *unstable.* A black hole therefore forms over the potential white hole and the resulting singularity gobbles up everything. Although detailed calculations have yet to be done, it seems reasonable to suppose that in a Penrose diagram of a *real* charged or rotating black hole, a *spacelike* singularity forms which cuts off all the future universes. Questions still remain as to how fast this occurs. The answer must depend on how rapidly light accumulates in a blue sheet along an event horizon exposed to $\mathscr{I}^-$ of some particular universe. If physicists who like the idea of white holes try to argue that the inevitable instability caused by the blue sheet occurs slowly, they find that they face a new difficulty of a very different kind, one that involves matter and antimatter.

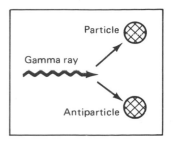

FIGURE 14-10. *Pair Creation.* Particles and antiparticles (for example, electron and positron, or proton and antiproton) can be created from high-energy gamma rays. Matter and antimatter are always created in equal amounts.

The existence of antimatter has been known to scientists for many years. Antimatter was first discovered in cosmic ray showers, and today antiparticles of all types are regularly produced in laboratory experiments involving nuclear physics. Nuclear physicists can create matter and antimatter most easily from high-energy gamma rays. As shown in Figure 14-10, under appropriate conditions a gamma ray can spontaneously turn into a particle and an antiparticle. This process occurs only if the gamma ray possesses a great deal of energy, more energy than is contained in the matter of the particles it creates. There is nothing mysterious about antimatter. Particles and antiparticles are always produced in equal numbers in the process called *pair creation.*

In examining pair creation, theoretical physicists have found it very useful to conceive of empty space as filled with imaginary or *virtual* pairs of particles. For example, a point in empty space may be thought of as being an imaginary electron sitting on top of an imaginary positron. Another point may be thought of as being an imaginary proton on top of an imaginary antiproton. In any such case the effects of the virtual particle are completely canceled by the effects of the virtual antiparticle. But, if an intense gamma ray comes along and strikes the virtual pair, the imaginary particles can absorb so much energy from the gamma ray that the radiation energy is converted into matter according to the famous equation $E = mc^2$ and the particles appear in the real world. The process of pair creation can therefore be understood in terms of virtual pairs of particles absorbing energy that causes them to materialize. The idea that empty space consists of virtual pairs waiting to be materialized is very useful in nuclear physics.

Think for a moment about conditions near a space-time singularity in a black hole. At a singularity, there is infinite warping of space-

245

time which gives rise to infinite tidal stresses. Anything falling into the singularity is completely torn apart by these overwhelming stresses. Therefore, very near the singularity, the tidal forces must be enormous. It is always possible to conceive of a place near a singularity where the tidal forces are large enough to pull apart anything. Specifically, consider empty space a fraction of a millimeter above a singularity. Although this space is empty, it can be thought of as containing virtual pairs of particles and antiparticles. Very close to the singularity, the tidal forces are so great that *virtual pairs are torn apart*. Gravity is so strong that imaginary electrons are torn away from imaginary positrons; imaginary protons are torn away from imaginary antiprotons. Calculations reveal that this process of tearing apart virtual pairs is so violent that each imaginary particle receives so much energy that it can materialize! Quite literally, the tidal forces of infinitely warped space-time at the singularity *tear apart space-time,* thereby producing matter and antimatter. Matter and antimatter therefore gush out of a singularity! Just as an intense gamma ray produces particles and antiparticles, the intensity of gravitation near a singularity produces particles and antiparticles.

As long as the singularity is spacelike and lies in the future, the particles and antiparticles cannot escape. But, as shown in Figure 14-11, if the singularity is timelike or lies in the past, then the matter and antimatter *can* escape. There are timelike paths along which the created material *can* escape.

The discovery of this remarkable process, due primarily to the brilliant theoretical work of Stephen W. Hawking, has some important implications. In a Schwarzschild hole, *if* the past singularity could exist, it would soon tear apart space-time in its vicinity and fill up the hole with matter and antimatter. In Reissner-Nordstrøm or

FIGURE 14-11. *Pair Creation near a Singularity.* Enormous tidal forces near a singularity tear apart space-time, thereby producing pairs of particles and antiparticles. This material can escape from the singularity if the singularity is timelike or is spacelike and lies in the past.

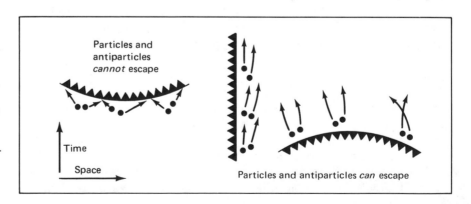

Particles and antiparticles *cannot* escape

Time

Space

Particles and antiparticles *can* escape

246

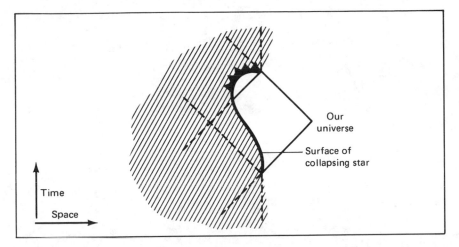

Our universe

Surface of collapsing star

Time

Space

FIGURE 14-12. *The Penrose Diagram of a Real Black Hole.* Pair creation by the singularity and the instability of blue sheets combine to choke off most of an idealized Penrose diagram. In all probability, a future singularity forms which is generally spacelike and eliminates any possibility of time machines or space travel to other universes.

Kerr holes, where there are timelike singularities, newly created pairs can also escape from the singularities. This same process of pair creation therefore rapidly fills up the Penrose diagram of charged and rotating holes.

So there is no hope! To permit travel to other universes, a hole must have timelike singularities. For there to be white holes, the singularities must lie somewhere in the past. In all such cases, these singularities produce enormous amounts of matter and antimatter which rapidly fill up the holes thereby eliminating any extremely strange processes. The instability of blue sheets and pair creation near the singularity choke off most of the Penrose diagram. Although detailed calculations have yet to be done, the Penrose diagram for a *real* rotating or charged black hole probably looks similar to Figure 14-12.

These theoretical developments of the mid-1970s may come as a severe disappointment to anyone who likes to fantasize about time machines and space travel to other universes. Often, however, a discovery in science that destroys one set of ideas gives rise to totally new concepts. Indeed, as we shall see in the final chapter, the process of pair creation by intense gravitational fields means that certain kinds of black holes can evaporate and explode!

15 GRAVITATIONAL WAVES AND GRAVITATIONAL LENSES

The general theory of relativity is by far the best description of gravity known to physicists today. This theory simply states that the gravitational field of any object manifests itself by warping space-time. Since all matter produces gravity, every material object exerts some distortion of the geometry of space-time. The more intense the gravitational field, the greater the distortion of space-time.

Imagine some objects moving around in space. Since each object has mass, each object carries a slight warping of space-time. As the objects move around, the geometry of space-time therefore changes. The geometry of space-time must readjust itself to the new configuration of matter every time the objects move. These readjustments appear as ripples in the overall geometry of space-time. The ripples move outward from the sources of gravity at the speed of light and are called *gravitational waves*.

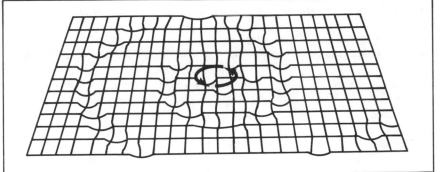

FIGURE 15-1. *Gravitational Waves from a Binary Star.* Two stars orbiting each other emit gravitational waves, as represented in this diagram. If the stars are very close together and orbit each other very rapidly, considerable amounts of power can be emitted in gravitational waves.

Gravitational waves are produced almost anytime matter moves around. A bouncing ball, a person waving his or her arms, and the moon orbiting the earth all produce gravitational waves. In general, objects with strong gravitational fields are capable of producing more intense gravitational waves than objects with weak gravitational fields. Two black holes orbiting each other in a hypothetical binary produce much more intense gravitational radiation than Jupiter orbiting the sun in our solar system. Table 15-1, based on calculations by Bragnisky, Ruffini, and Wheeler, gives the power emitted by various types of binary systems. For comparison, the sun's energy output of light is 400,000 quadrillion megawatts.

From Table 15-1 you can see that short-period binaries in general emit more powerful gravitational waves than long-period binaries. In short-period binaries, the stars are very close together and their gravitational interaction is very strong. See Figure 15-1.

TABLE 15-1   Astronomical Sources of Gravitational Waves

Type of Binary	Name of Binary	Orbital Period	Power Emitted, in Gravitational Waves
Sun and Jupiter	Solar system	11.9 years	5.2 kilowatts
Binary star	η Cas	480 years	5.6 kilowatts
Binary star	ξ Boo	150 years	360 kilowatts
Binary star	Sirius	50 years	110 megawatts
Binary star	β Lyr	13 days	6 quadrillion megawatts
Binary star	UV Leo	14 hours	63 quadrillion megawatts

TABLE 15-2   Hypothetical Sources of Gravitational Waves

Separation between Two Collapsed Stars	Orbital Period	Power Emitted, in Gravitational Waves
6,200 miles	12 seconds	30 trillion quadrillion megawatts
600 miles	$\frac{1}{3}$ second	3 quadrillion quintillion megawatts
60 miles	$\frac{1}{100}$ second	3 quadrillion quadrillion quintillion megawatts

The strongest gravitational waves would be emitted by two neutron stars or two black holes orbiting each other. Table 15-2 gives the power emitted by two such dead stars. This table, adapted from calculations by Ruffini and Wheeler, is for two 1-solar-mass neutron stars (or two 1-solar-mass black holes) in very close orbit about each other.

Although some of the binaries listed in Tables 15-1 and 15-2 do emit considerable power in the form of gravitational waves, the amount of energy that finally arrives at the earth can be quite small. The farther away the source is, the weaker it appears. The physicist denotes the detectable strength of a source of radiation in terms of how many watts per square foot (or per square mile) actually strike the earth. For one of the most powerful sources of gravitational radiation, the binary star UV Leo, which is 220 light-years away, the *flux* of energy at the earth is only $\frac{1}{3}$ billionths of a watt per square mile. That is weak!

Assuming the existence of black hole binaries, the flux at the earth can be considerably higher. For example, for a binary consisting of two black holes of one solar mass each, separated by 60 miles (orbital period $= \frac{1}{100}$ second), and at a distance of 3,260 light-years, the flux at the earth is about 17 watts per square inch. For comparison, the energy from the sun striking the earth is about 1 watt per square inch. Nevertheless, most astronomers would argue vigorously that close black hole binaries are very unlikely in the first place.

Mathematically, the possible existence of gravitational waves has been known for many years. Physically detecting gravitational radiation in laboratory experiments, however, turns out to be a very difficult task. Compared to other types of radiation, gravitational waves are incredibly weak. For example, the electromagnetic radiation—

such as radio waves—emitted by an oscillating electric charge is a trillion trillion trillion times stronger than the gravitational waves emitted by this same charge. The primary reason is that electromagnetic forces are very much stronger than gravitational forces. While it is therefore easy to detect electromagnetic radiation using a variety of devices such as the human eye, photographic film, or a radio, building machines that respond to gravitational radiation poses some enormous problems for the experimental physicist.

The human eye or a piece of photographic film responds to electromagnetic radiation of a particular wavelength because the oscillating electric and magnetic fields of the radiation set charged particles in motion. When the human eye is exposed to light, the resulting vibrations of electrons in the atoms of cells in the retina produce a weak electric current that is transmitted to the brain. In the case of photographic film, the motions of the electrons in the atoms of the emulsion initiate chemical reactions. In general, electromagnetic radiation causes charged particles to move, and the radiation is detectable because of the motions of these particles.

Gravitational waves also cause particles to move. However, gravitational waves give rise to a very different kind of motion than do electromagnetic waves. To compare the effects of these two different kinds of waves, imagine a ring of electrons floating in free space. If an electromagnetic wave passes through the ring, all the electrons oscillate back and forth in unison. But if a gravitational wave passes through the ring, the particles are set in motion relative to each other. As shown in Figure 15-2, the shape of the ring is distorted. The ring flexes.

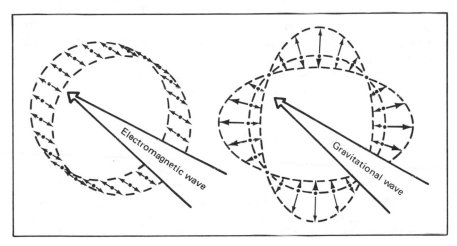

FIGURE 15-2. *Electromagnetic versus Gravitational Waves.* When an electromagnetic wave passes through a ring of electrons floating in space, all the electrons oscillate back and forth in unison. If a gravitational wave passes through a ring of particles, the shape of the ring is distorted because the particles are set in motion relative to each other.

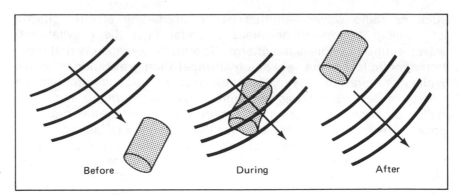

FIGURE 15-3. *Gravitational Waves Striking an Object.* As gravitational waves move through an object, the shape of the object is distorted very slightly.

Before   During   After

This distorting effect of gravitational waves suggests a method by which gravitational radiation might be detected. Imagine a large object such as a metal cylinder, as shown in Figure 15-3. In the absence of gravitational radiation, the cylinder has a certain shape. However, if a gravitational wave passes through the cylinder, it will flex. The shape of the cylinder will be distorted very slightly in response to the wave. *If* these tiny distortions could be detected, physicists would have a means of building a *gravitational wave antenna.*

The task of building a gravitational wave antenna was first tackled by Joseph Weber at the University of Maryland (see Figure 15-4). Beginning in the 1950s, Weber decided to work with a large aluminum cylinder. Realizing that he must be able to detect incredibly minute changes in the shape of the cylinder, he glued piezoelectric strain transducers to the surface of the bar. These piezoelectric crystals are extremely sensitive and produce a small electric current when subjected to even the tiniest pressure or stress. The resulting electric signal is amplified and recorded by electronic apparatus in Weber's laboratory.

In principle, the basic idea is very simple. Every time a gravitational wave passes through Weber's cylinder, it flexes. The vibrations of the bar shake the piezoelectric crystals, thereby producing an electric current that is amplified and recorded. In practice, the experiment is extremely difficult. A thunderstorm or a person walking down the stairs in the laboratory produces enough vibration to trigger the piezoelectric crystals. Weber was therefore faced with the problem of discriminating between true gravitational waves and "noise" due to extraneous effects.

In the 1960s, Weber set up *two* gravitational wave antennas, one at the University of Maryland near Washington, D.C., and the other

FIGURE 15-4. *A Gravitational Wave Antenna.* Weber is shown here standing alongside one of his antennas. The antenna consists of a large aluminum cylinder whose vibrations are detected by means of very sensitive crystals glued to the cylinder. (Courtesy of Joseph Weber)

at Argonne National Laboratory outside of Chicago. Each cylinder was about five feet long and two feet in diameter, and weighed about 3,000 pounds. As expected, the piezoelectric crystals on each bar were constantly putting out signals as the bars vibrated in response to all sorts of random effects. However, Weber noted that it would be quite unlikely that two random effects should occur simultaneously in both Washington and Chicago. He therefore focused his attention only on *coincidences.* If a gravitational wave strikes the earth, both of his antennas should vibrate in essentially the same way at the same time. By ignoring random vibrations, which do *not* occur simultaneously at Maryland and Argonne, he should be able to eliminate the "noise" (see Figure 15-5).

After many years of diligent work, Weber finally succeeded in detecting simultaneous oscillations of both his antennas. By 1968, the apparatus had been perfected to the extent that coincidences in the signals from the antennas at Maryland and Chicago were observed on a daily basis. Weber presented arguments that the probability of these coincidences being due to chance was extremely small and therefore concluded that he had, in fact, detected gravitational waves. In addition, the antennas used in this experiment have a crude directional sensitivity. They are more sensitive to gravitational waves coming from certain directions in the sky. Noting the time

253

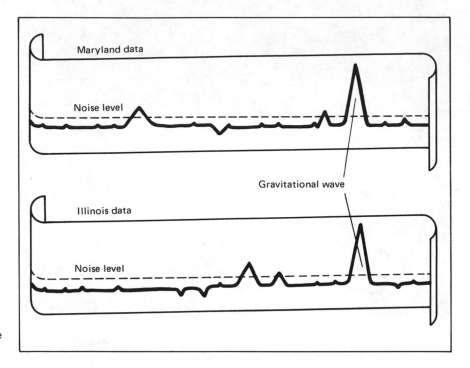

FIGURE 15-5. *The Detection of a Gravitational Wave.* Gravitational wave antennas at the University of Maryland and Argonne National Laboratory are constantly putting out signals originating from random effects. However, a strong signal observed *simultaneously* by the two antennas may be due to a gravitational wave.

of the day during which most of the coincidences were observed, Weber suggested that gravitational waves might be coming from the center of our galaxy.

The claim that Weber has indeed detected gravitational waves is hotly contested by many physicists. There are primarily two reasons for the debate. First of all, in recent years several other teams of physicists have built gravitational wave antennas but have detected virtually nothing. This means that either Weber is wrong and his co-incidences are due to chance or some other effect, or the physicists who are constructing new antennas have yet to improve their apparatus to the extent that they can detect the same waves that Weber reports observing.

The second point of contention comes from some interesting calculations. We have noted that gravitational waves are *very* weak. Compared to other forms of radiation, they carry very little energy. This, of course, is why gravitational waves are so very difficult to detect. But *if* Weber is in fact observing gravitational waves from the center of our galaxy, it is possible to estimate the amount of energy

254

that is necessary to produce the waves in the first place. The kinds of processes that could produce substantial amounts of gravitational radiation include wildly vibrating neutron stars or black holes, debris falling into a black hole, collisions between two black holes, and supernova explosions. In order for these or similar processes to produce the power required to activate Weber's apparatus, incredibly violent events must be occurring at the center of our galaxy. Furthermore, since Weber claims to detect several coincidences every day, these incredibly violent events must be taking place *hourly.* Astrophysicists find it virtually impossible to conceive of circumstances under which processes could produce so much gravitational radiation so frequently.

Although it seems inconceivable that supernovas could be exploding almost hourly or that black holes collide several times a day at the center of the galaxy, perhaps there are some avenues which have yet to be explored. At first glance, the processes that might give rise to Weber's gravitational waves emit energy in all directions. Since the earth is so tiny, it intercepts only a very small fraction of the total energy output of these processes. That is why astrophysicists generally believe the processes must be so incredibly violent. But if the released energy is somehow *beamed* toward the earth, far less violent events could be used to explain Weber's observations.

Earlier in this chapter, it was noted that black holes in binary systems could conceivably emit much more intense gravitational radiation than ordinary stars in orbit about each other. Although it might be virtually impossible to find a black hole binary, perhaps black holes sometimes pass very close to each other as they move around the galaxy. Indeed, the center of our galaxy is perhaps the most likely place for close encounters between black holes since there are more stars at the galactic center than anywhere else in the Milky Way. Each time two black holes pass within a few miles of each other, their interaction will produce a strong burst of gravitational waves. More important, the resulting gravitational waves will be *focused.* Just as a black hole bends light rays, it will also deflect gravitational waves. Specifically, a black hole can focus radiation (either electromagnetic or gravitational) in certain directions so that a distant observer might see an unusually intense burst of energy. This focusing property of a black hole is called a *gravitational lens.*

In the mid-1970s, Thorne and Kovacs at Caltech began examining the possibility of gravitational focusing of the gravitational waves emitted by near-collisions of black holes. Although their work is not yet complete, the idea that gravitational waves occasionally are beamed toward the earth by these near-collisions seems to offer great promise. Perhaps this beaming is the mechanism that gives rise to the waves Weber is detecting with his antennas.

The concept of gravitational lenses, while possibly relevant to the beaming of gravitational waves, is an inherently interesting topic. Ever since black holes were first proposed, it was realized that the curvature of space-time around a black hole would severely deflect light. Under appropriate circumstances, beams of light approaching a black hole could be bent in such a way that distant objects might appear brighter than usual or perhaps multiple images would be seen. As shown in Figure 15-6, a black hole located between an observer and a distant source of light focuses beams of light. In theory, a gravitational lens will produce an infinite number of images of distant stars and galaxies. In practice, only the *primary* and *secondary images* will contain enough light to be seen.

Since the detection of multiple images of the same object would constitute the conclusive discovery of a black hole, many astrophysicists have performed laborious calculations concerning the brightnesses and shapes of images formed by gravitational lenses. For example, in 1974, R. C. Wayte at Imperial College, London, published drawings of what a galaxy should look like when seen "through" a gravitational lens. The undistorted image of a typical galaxy is shown at the top of Figure 15-7. If a black hole lies between the earth and a distant galaxy, two images would be seen. In addition, the two resulting images are substantially distorted. The closer the black hole is to a straight line directly joining the earth and the galaxy, the more the images are distorted.

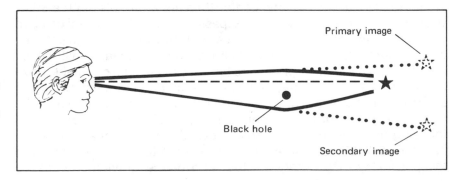

FIGURE 15-6. *A Gravitational Lens*. A black hole deflects and focuses light rays from a distant star. Most of the deflected light is concentrated into two images of the distant star.

256

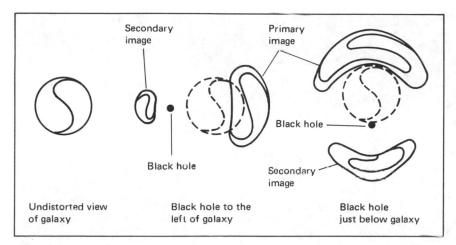

Secondary image

Primary image

Black hole

Black hole

Secondary image

Undistorted view of galaxy

Black hole to the left of galaxy

Black hole just below galaxy

FIGURE 15-7. *A Galaxy Seen "Through" a Gravitational Lens.* The undistorted view of a typical spiral galaxy is shown at the left of the diagram. If a black hole is located between the earth and the galaxy, two images of the distant galaxy would be seen by astronomers. The closer the black hole is to a straight line joining the earth and the galaxy, the more the resulting images are distorted. (Adapted from R. C. Wayte.)

Astronomers have never seen images of galaxies that look like the drawings produced by Wayte. Perhaps the primary reason is that the alignment between the earth, the black hole, and the distant galaxy must be extremely precise in order to cause any substantial gravitational focusing. The probability of a near-perfect alignment is exceedingly small.

Although gravitational lenses have yet to be discovered, an important conclusion can be drawn. With the aid of powerful telescopes, astronomers see numerous galaxies scattered all over the sky. Galaxies tend to group together in *clusters* which are seen in every unobscured part of the sky (see Figure 15-8). If *supermassive black holes* exist, if black holes containing a billion solar masses exist in the universe, they should severely distort the overall appearance of the sky. A supermassive black hole somewhere in the universe would have a substantial effect on the images of background galaxies. Since clusters of galaxies are scattered rather uniformly around the sky, it is possible to conclude that supermassive black holes do *not* exist.

A likely place where gravitational lenses would be important is in binary systems containing a black hole and a visible star. The black hole will deflect and focus the light from the visible star, thereby producing some unusual effects. In the early 1970s, C. T. Cunningham and J. M. Bardeen performed some interesting calculations that show the appearance of the images of the visible star in such a binary.

257

FIGURE 15-8. *A Cluster of Galaxies*. Numerous galaxies are seen in every unobscured region of the sky. A super-massive black hole some-where in the universe would severely distort the images of galaxies over a large part of the sky. Such distortions have never been observed. (Hale Observatories)

In the absence of any relativistic effects, the orbit of one star about another in a binary should look like an ellipse, as shown in Figure 15-9. But if one of the members of the binary is a black hole, it will have substantial influence on the locations of images of the visible star. Figure 15-10 shows the path of the *primary image* of the visible star as it orbits an extreme Kerr black hole ($M = a$). When the visible star is in front of the black hole, the star's image is virtually unaffected by the black hole. But as the star swings behind the black hole, most of the light rays that could have otherwise been seen by earth-based astronomers are now swallowed by the hole. Only those light rays emitted from the visible star at very high angles manage to get past the hole's intense gravitational field. The appar-ent path of the visible star when it is behind the black hole is there-fore severely distorted.

258

In addition to the primary image discussed above, secondary images could be seen by earth-based astronomers. The path of the secondary image of the visible star is shown in Figure 15-11. For comparison, the path of the primary image is shown with dashed lines. In the case of the secondary images, the light rays from the visible star orbit once around the black hole before escaping to the distant observer. Although comparatively unimportant for the primary image, the rotation of the black hole strongly influences the location of the secondary images. When the visible star is actually in front of the black hole, the secondary image appears to the left of the black hole. As the star moves around the black hole (counterclockwise as seen against the sky), the secondary image also traces out a path counterclockwise around the location of the black hole. However, about three-fourths the way through the orbit, when the visible star is coming around from behind the black hole, something quite unusual happens. A new secondary image is seen. This new image splits in two. One half of it moves counterclockwise, completing the orbit, while the other half moves clockwise and meets the initial secondary image. Multiple secondary images occur because the black hole is rotating. The path one way around the black hole is different from the path the other way. The differences between counterrotating and corotating light paths around the Kerr black hole give rise to several secondary images.

Of course, there are higher-order images of the visible star. Light rays that orbit the black hole two, three, or four times will each produce a complicated set of images. But most of the light from the

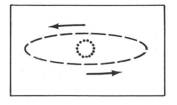

FIGURE 15-9. *An Ordinary Binary.* Neglecting relativistic effects, the orbit of one star about another in a binary system is an ellipse.

FIGURE 15-10. *The Path of the Primary Image.* The location of the primary image of a visible star orbiting a rotating black hole is shown in this sequence of drawings. When the visible star is in front of the black hole, the location of the image is only slightly affected. But as the visible star passes behind the black hole, only those light rays leaving the star at high angles manage to pass by the black hole. (Adapted from C. T. Cunningham and J. M. Bardeen.)

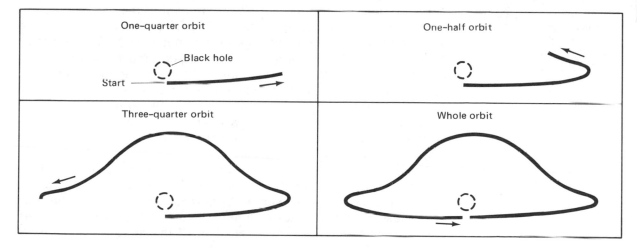

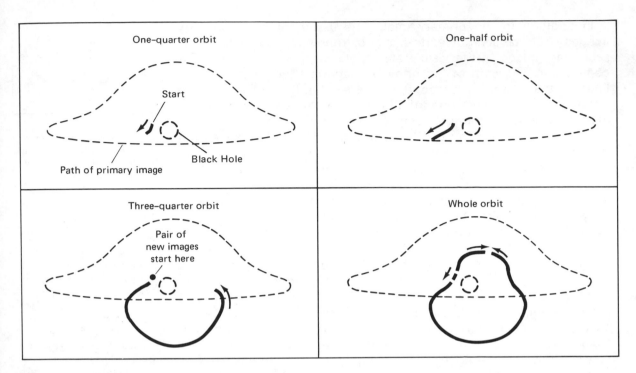

One-quarter orbit

Start

Black Hole

Path of primary image

One-half orbit

Three-quarter orbit

Pair of
new images
start here

Whole orbit

FIGURE 15-11. *The Path of the Secondary Images.* Light that orbits once around the black hole before escaping to a distant astronomer produces secondary images. There are several secondary images because the black hole is rotating. (Adapted from C. T. Cunningham and J. M. Bardeen.)

visible star is concentrated in the primary and secondary images alone. The higher-order images would be exceedingly weak.

In order for the gravitational focusing of a black hole in a binary to be important, the black hole and the visible star must appear close to each other in the sky. The alignment of the earth, the black hole, and the visible star must be nearly perfect. But if the black hole and the visible star are close together, then it is very unlikely that an astronomer could ever see individual images moving around. The images would be so close together and their motions would be so tiny that even with the most powerful telescopes the entire system would look like a single stationary blob. It is therefore quite improbable that astronomers will ever discover a black hole from the peculiar motions of a visible star in a binary. However, there is a second effect that does hold some hope.

The previous discussion of a visible star orbiting a black hole treated only the locations of the various images. During the course of an orbit, these images vary in brightness. In examining the total brightness of all the images, Cunningham and Bardeen discovered that at regular intervals the intensities of the images combine. At

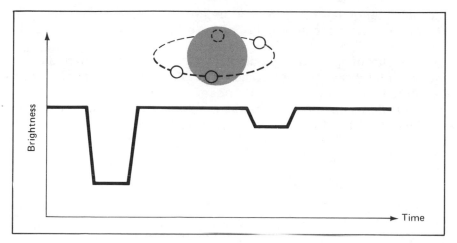

least once during each orbit, the focusing of the light from the visible star is unusually powerful and the visible star momentarily appears much brighter than normal.

Astronomers are very familiar with binary stars that periodically brighten and dim. If a binary is oriented in space so that one star alternately passes in front of the other, eclipses occur. During eclipses, one star obscures some of the light from the other star. The light from an *eclipsing binary* periodically diminishes, as indicated in Figure 15-12, each time an eclipse occurs. Even though astronomers might not be able to distinguish two separate stars

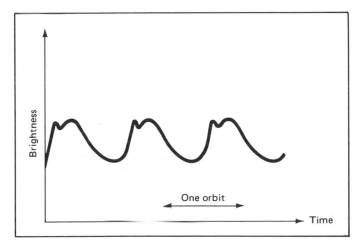

FIGURE 15-13. *The Light Curve of a Binary Containing a Black Hole.* The focusing effect of a black hole in a binary should produce spikes in the binary's light curve.

261

through their telescopes, by observing periodic dips in the *light curve* they have conclusive proof of the existence of an eclipsing binary.

If one star in an eclipsing binary is a black hole, it will periodically focus the light rays from the visible star. For a brief moment during each orbit, the visible star will appear unusually bright. The light curve of such a binary will therefore exhibit regular spikes, as shown in Figure 15-13. If the light curve of a binary were ever discovered with anomalous spikes, it would seem highly probable that the binary contains a black hole.

16 EXPLODING GALAXIES AND MASSIVE BLACK HOLES

One of the great fads in astronomy during the eighteenth and nineteenth centuries was comet-hunting. Serious astronomers in Western Europe spent many long nights scanning the skies in search for new comets. Prizes and medals were awarded to those successful astronomers who managed to find one or more of these ephemeral members of the solar system.

A newly discovered comet often appears as a faint fuzzy blob through the astronomer's telescope. Only as the comet swings near the sun are the comet's ices vaporized, thereby producing a characteristic tail. By examining the path of a newly discovered comet against the background of fixed stars from night to night, it is possible to calculate the comet's orbit about the sun. If the orbit happens to take the comet near the sun's surface, a substantial fraction of the cometary ices will be vaporized. This produces a spectacular sight often readily visible to the naked eye. Usually,

FIGURE 16-1. *A Galaxy.* Objects like this spiral galaxy (M 101) were included in Messier's catalog. What the distances to galaxies are was hotly debated by astronomers during the early 1920s. (Lick Observatory)

however, the orbit is so far from the sun that the comet always remains dim and can be seen only with the aid of a telescope.

One of the problems that faced comet hunters was that there are many objects in the sky that look like comets at first glance. Through a telescope they appear as small, fuzzy objects, but observations spanning many nights reveal that they do not change their positions with respect to the background stars. This fact indicates that these fuzzy objects, or *nebulas,* are very far from the solar system.

Almost two centuries ago, the famous French comet hunter Charles Messier compiled a list of about 100 nebulas. Although his initial goal was to aid his colleagues by identifying objects that should not be mistaken for comets, his catalog was soon realized to contain some of the most interesting objects in the sky. The first object in his catalog (M 1) is the Crab nebula shown in Figure 7-5. The forty-second object (M 42) is the Orion nebula, and the fifty-seventh object (M 57) is the ring nebula; these are shown in Figures 6-6 and 6-9, respectively.

Many of the objects listed in the *Messier Catalogue* are relatively nearby. Like the Orion nebula, the Crab nebula, and the ring nebula, they are often related to the births and deaths of stars. They are located among the stars and therefore are at typical stellar distances. These objects are scattered around our own Milky Way Galaxy.

Although many of the Messier objects are true clouds of gas or clusters of stars floating in interstellar space, several dozen objects in this famous catalog are definitely very different from all the others. These unusual objects frequently have a pinwheel shape and were therefore called *spiral nebulas.* M 101, shown in Figure 16-1, is a good example. These spiral nebulas were destined to be the subject of one of the greatest debates in astronomy since the time of Copernicus.

The central issue involving the spiral nebulas was their distances. Are they relatively nearby objects only a few hundred or thousand light-years away? Or are they incredibly huge objects millions of light-years from the Milky Way? By the early 1920s, professional astronomers were divided on this subject and heated debates raged at every scientific meeting.

The issue was finally settled in the mid-1920s when a young astronomer at Mount Wilson Observatory, Edwin Hubble, announced his discovery of variable stars in several spiral nebulas such as M 31. M 31 is a large spiral nebula in the constellation of Andromeda. A portion of this object is shown in Figure 16-2, with two variable stars

FIGURE 16-2. *Variable Stars in M 31.* This photograph of part of the Andromeda galaxy (M 31, also called NGC 224) shows two variable stars (marked with white lines). From studying variable stars, Hubble was able to prove that the galaxies were very far away from our own Milky Way Galaxy. (Hale Observatories)

identified. The discovery of these variables proved invaluable because their period of variability is directly related to their real brightnesses. By measuring the frequency with which these stars changed their luminosity, Hubble was able to deduce their distance. The distance to M 31 is about two million light-years. This is very far beyond the confines of our own Milky Way Galaxy and clearly indicates that the spiral nebulas are huge stellar systems. These vast stellar systems are today called *spiral galaxies.*

During the late 1920s, Hubble continued to make important discoveries concerning the nature of extragalactic objects. First of all, he realized that galaxies can be classified into four basic types. In addition to ordinary spiral galaxies such as M 31 and M 101, there are similar objects that have a bar extending through their nuclei. The spiral arms of such galaxies originate from the ends of the bar rather than from the nucleus. Figure 1-4 shows a good example of one of these *barred spiral galaxies.*

In addition to spiral and barred spiral galaxies, there is a third class of objects that exhibit no spiral arms at all. Such galaxies have an elliptical, amorphous appearance and are called *elliptical galaxies.* Some of the brightest galaxies in the sky, such as M 87 in Figure 16-3, are ellipticals.

FIGURE 16-3. *The Giant Elliptical Galaxy M 87.* Many galaxies, such as this object in the constellation of Virgo, show no spiral arms at all and are called elliptical galaxies. Some of the brightest galaxies in the sky are ellipticals. (Lick Observatory)

Finally, galaxies that do not fit into one of the three previous classifications (spirals, barred spirals, or ellipticals) often have unusual shapes. These strange-looking galaxies, such as M 82 in Figure 16-4, are called *irregulars*. The two galaxies nearest our own Milky Way, the Small and Large Magellanic Clouds, which can be seen with the naked eye from southern latitudes, are also irregulars.

In addition to inventing a classification scheme for galaxies based on their appearance, Hubble also took spectra of many extragalactic objects. In examining the spectra of galaxies he noticed that virtually all such objects exhibit *redshifts*. As shown in Figure 16-5, their spectral lines were shifted toward the red end of the spectrum. By 1929, Hubble realized that the redshifts of galaxies were directly correlated with their distances. Nearby galaxies have low redshifts and therefore are moving away from us slowly. More distant galaxies have much higher redshifts and therefore are moving away from us much more rapidly. This relationship between distance and reces-

FIGURE 16-4. *The Exploding Galaxy M 82.* This object in the constellation Ursa Major is a good example of an irregular galaxy. The distorted appearance of M 82 occurs because the galaxy is in the process of exploding. (Hale Observatories)

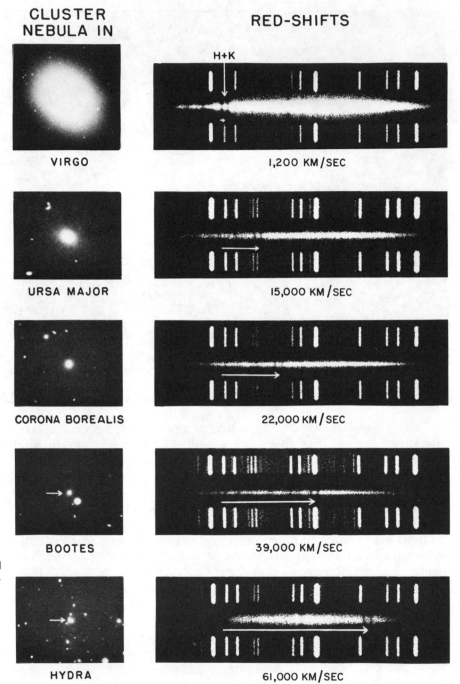

CLUSTER
NEBULA IN

RED-SHIFTS

H+K

VIRGO

1,200 KM/SEC

URSA MAJOR

15,000 KM/SEC

CORONA BOREALIS

22,000 KM/SEC

BOOTES

39,000 KM/SEC

HYDRA

61,000 KM/SEC

FIGURE 16-5. *The Redshifts of Galaxies.* Photographs and spectra of five elliptical galaxies. They all exhibit redshifts that are directly correlated with their distances. This fact reveals that the universe is expanding. (Hale Observatories)

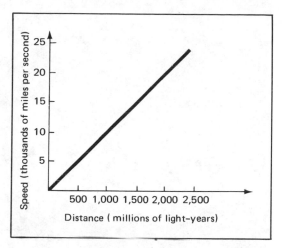

FIGURE 16-6. *The Hubble Law*. The distances and speeds of galaxies are directly related. Nearby galaxies are moving away from us very slowly and more distant galaxies are rushing away from us at much higher speeds. This relationship reveals that the universe is expanding.

sional velocity is best displayed in the form of a graph as shown in Figure 16-6 and is called the *Hubble law.*

The primary significance of the Hubble law is that since all galaxies are getting farther and farther apart, *the universe must be expanding!* Furthermore, from the orientation of the line in Figure 16-6 it is possible to calculate when all the galaxies must have been piled on top of each other. According to the best data available today, primarily as a result of the work of Allan Sandage at Hale Observatories, the universe must have been in a highly condensed state 18 billion years ago. Then, for some unknown reason, there must have been a stupendous explosion, the *big bang,* throughout all space, which started the universe expanding. The most straightforward conclusions one can draw from Hubble's pioneering work is that the universe is expanding as the result of a primordial explosion roughly 18 billion years ago. These are the general features of the *big bang cosmology.*

In the 1960s, astronomers began making very detailed observations of galaxies, and much to their surprise found that incredibly violent events often seem to be occurring at galactic centers. For example, the irregular galaxy M 82 is being rocked by an incredible explosion that is literally tearing the galaxy apart. Jets of gas are surging up out of the galaxy at speeds in excess of 10,000 miles per second.

In addition to irregular galaxies being deformed by violent events, ordinary-appearing galaxies were often found to be experiencing cataclysmic explosions. For example, a short time-exposure of the

FIGURE 16-7. *The Jet in M 87.* A short time-exposure of the elliptical galaxy M 87 (compare with Figure 16-3) reveals a huge jet of gas surging up out of the galaxy's nucleus. The jet is a powerful source of radio noise. (Lick Observatory)

FIGURE 16-8. *The Exploding Galaxy NGC 1275.* This remarkable photograph reveals great quantities of gas being ejected from NGC 1275 in the constellation of Perseus. (Courtesy of R. Lynds, Kitt Peak National Observatory)

270

giant elliptical galaxy M 87 reveals a huge jet of gas erupting from the galaxy's nucleus, as shown in Figure 16-7. Physicists find it almost impossible to explain the amounts of energy that must be involved in such explosions.

One of the most impressive examples of an exploding galaxy is NGC 1275 in the constellation of Perseus ("NGC" stands for *New General Catalogue,* which lists thousands of nebulas). Figure 16-8 shows a remarkable photograph of this object taken by R. Lynds at Kitt Peak National Observatory. NGC 1275 looks more like a supernova remnant such as the Crab nebula than a galaxy. It is also a powerful source of radio noise and x-rays.

One of the most powerful radio sources in the southern sky is associated with the unusual galaxy NGC 5128. As shown in Figure 16-9, this galaxy is also being torn apart by a stupendous explosion. The processes that could account for a galaxy blowing itself apart remain a total mystery to the modern scientist.

FIGURE 16-9. *The Exploding Galaxy NGC 5128.* The violent events tearing this galaxy apart also produce vast quantities of radio noise. NGC 5128 is the visible galaxy associated with one of the most powerful radio sources in the southern sky. (Hale Observatories)

FIGURE 16-10. *The Quasar 3C 48.* Quasars look like bluish stars but have enormous redshifts. From the Hubble law, their huge redshifts mean that they must be the most distant objects seen in the universe. Since they are so far away, they must be emitting vast quantities of energy, far more energy than is emitted by an ordinary galaxy. (Hale Observatories)

The theoretical problems of trying to understand the processes responsible for violent events in galactic nuclei have been further compounded by the discovery of quasars. Quasars such as 3C 48 shown in Figure 16-10 are bluish, star-like objects whose spectra exhibit enormous redshifts. Indeed, there are some quasars having redshifts corresponding to recessional velocities greater than 90 percent of the speed of light! Turning to the Hubble law, astronomers realize that these large redshifts mean that the quasars must be extremely far away. They are, in fact, so far away that by any reasonable standards they should be too dim to be seen at all. As with the case of exploding galaxies, quasars seem to be emitting fantastic amounts of energy whose origin appears to defy explanation.

In 1969, D. Lynden-Bell of the Royal Greenwich Observatory undertook an interesting study of quasars and exploding galaxies. On very general grounds he showed that whatever is responsible for the enormous energy-output at galactic nuclei or in quasars must be *small* and *massive.* Typically, the masses of these energy-producing objects could be as much as 10 million suns! Furthermore, since these objects must be small, it logically follows that either (1) they are already black holes, or (2) they will become black holes in a very short period of time. Lynden-Bell therefore concluded that *massive black holes* might be related to the mysteries surrounding exploding galaxies and quasars.

Although massive black holes might be located at the centers of

272

galaxies, it is not easy to understand exactly how they are producing the enormous energies astronomers observe. In 1971 Lynden-Bell and M. J. Rees of Cambridge University showed that a hypothetical massive black hole at the center of our galaxy should be surrounded by a huge accretion disk. Unlike Cygnus X-1, however, which emits x-rays, this accretion disk should emit vast quantities of infrared radiation. It is interesting to note that as a result of the work of astronomers such as Frank Low at the University of Arizona, powerful sources of infrared radiation have been discovered at the center of the Milky Way.

Another possibility concerning energy production by massive black holes involves the Penrose mechanism discussed in Chapter 11. Recall that if an object falling into the ergosphere of a Kerr black hole breaks apart, a piece of the object can be ejected back out into space with enormous energy. It is well known to astronomers that most galaxies contain *globular clusters.* These globular clusters, one of which is shown in Figure 16-11, typically consist of 100,000

FIGURE 16-11. *A Globular Cluster.* Globular clusters are found around almost every galaxy. Typically they contain about 100,000 stars and are in highly elliptical orbits about the nuclei of the galaxies to which they belong. (Lick Observatory)

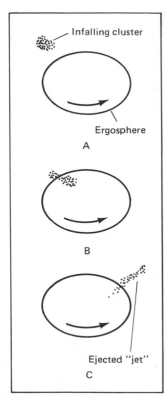

Infalling cluster

Ergosphere

A

B

Ejected "jet"

C

FIGURE 16-12. *Percolation through the Ergosphere.* A cluster of stars falling into the ergosphere surrounding a massive black hole would be torn apart. Through the Penrose mechanism, some of the stars could be ejected back out into space with enormous speeds.

stars and are in highly elliptical orbits about the center of the galaxy to which they belong. If there is a massive black hole at the center of a galaxy, it is conceivable that a globular cluster occasionally passes through the hole's ergosphere. Such an encounter would tear the cluster apart, as indicated by Figure 16-12. If some of the stars fall down into the hole, the remaining stars could be ejected violently back out into space. Perhaps this process, known as *percolation through the ergosphere,* is responsible for the "jets" seen in galaxies like M 87 (Figure 16-7) or the quasar 3C 273 (Figure 5-8).

Although the possibility of massive black holes at the centers of galaxies has some appealing features, it should be emphasized that there is no direct observational evidence to support this hypothesis. This does not seem to inhibit some astrophysicists from engaging in rather fanciful speculation. For example, even more speculative and intriguing is the idea that many of the processes observed in galaxies and quasars are associated with *white holes.* In Chapter 14 general arguments were presented which would appear to rule out the possibility of white holes. One of the primary motivations for these arguments centers about the fact that the existence of white holes would appear to violate certain principles in which modern scientists fervently believe, such as causality.

In 1975, Stephen Hawking at Caltech pointed out that all the laws of physics are formulated in terms of space and time. But at the singularity, where there is infinite curvature, concepts like "space" and "time" lose meaning. Therefore at the singularity the laws of physics also have no meaning. A space-time singularity does not have to obey the traditional laws of nature and a naked singularity could spew out matter and energy in a totally random fashion. This incredible discovery by Hawking, called the *randomicity principle,* means that a singularity looks like a white hole. An exposed singularity at the center of a galaxy could indeed account for the mysterious processes astronomers have discovered in recent years.

Back at the turn of the century, astronomers realized that they could not explain why the sun shines within the framework of traditional existing physics. By all reasonable standards, the sun should have stopped shining many millions of years ago. Only after the development of nuclear physics, special relativity, and quantum mechanics could the pieces be fitted together. The dilemma of the shining sun was resolved only after the discovery of new laws of physics.

Modern science again faces a serious dilemma. Almost daily astronomers discover objects in the sky emitting amounts of energy

274

that defy explanation. A few astrophysicists have turned to some of the most unusual objects known to science: massive black holes, white holes, and naked singularities. Perhaps they are on the right track. Or perhaps their work is just a desperate, futile effort. The history of astronomy would suggest that we should not rule out the possibility of new laws of physics. Perhaps only after a new and deeper understanding of physical reality has been discovered will astronomers finally understand the nature of quasars and exploding galaxies.

17 PRIMORDIAL BLACK HOLES

During the early years of the space program it was realized that worldwide communication networks could be one of the practical applications of space exploration. With earth-orbiting satellites as relay stations, everything from telephone calls to television programs could be beamed to any location on our planet. At both governmental and private laboratories, scientists and engineers therefore turned their attention to developing the technology necessary to realize this particular benefit of the conquest of space.

In the mid-1960s, scientists at Bell Telephone Laboratories in New Jersey began constructing a very sensitive antenna with which to receive the weak signals from Echo and Telstar communication satellites. In 1965, Robert W. Wilson and Arno A. Penzias performed a series of experiments with the Bell Telephone antenna to measure and account for all sources of static which normally accompany electronic equipment and radio receivers

FIGURE 17-1. *Wilson and Penzias.* Robert W. Wilson (left) and Arno A. Penzias (right) are shown standing in front of the horn-reflector antenna at Bell Telephone Laboratories in Holmdel, New Jersey. Using this antenna, Wilson and Penzias discovered the 3° background radiation that is thought to be the cooled-off "echo" of the big bang. (Courtesy of Bell Laboratories.)

(see Figure 17-1). Much to their surprise, there was a small amount of noise that they could *not* account for or explain. Even more surprising is the fact that this weak static seemed to be coming from all parts of the sky with equal intensity.

Unknown to the Bell Telephone scientists, a team of physicists including R. H. Dicke, P. J. E. Peebles, P. G. Roll, and D. T. Wilkinson were making some interesting theoretical calculations a few miles away at Princeton University. Following earlier suggestions by Tolman and Gamow, the Princeton team was exploring some of the consequences of living in a big-bang universe. If the universe began with a primordial explosion 18 billion years ago, the temperature of the early universe must have been extremely high. For example, one second after the creation, the temperature of the universe was probably about 10 billion degrees. The Princeton team further realized that a basic consequence of thermodynamics is that when anything such as a gas is allowed to expand, its temperature decreases. Since the universe is expanding, the temperature of the universe must be decreasing. For example, a couple of hours after the creation event, the temperature of the universe was probably down to a hundred million degrees. When the universe was one century old, the temperature had decreased to slightly under one million degrees. Indeed, today, 18 billion years after the *primordial fireball,* the temperature should be about 3° above absolute zero, the lowest possible temperature.

One of the basic laws of physics relates the temperatures of objects to the kinds of radiation they emit. For example, an object whose temperature is about 6000 °K (for example, the sun's surface) emits primarily visible light. Extremely hot objects with temperatures of a million degrees emit x-rays. Very cool objects emit radio waves. Specifically, anything at 3 °K should emit weak radio waves primarily at wavelengths ranging from 1 millimeter to about 100 centimeters. The Bell Telephone antenna was "tuned" to 7.53 centimeters, almost exactly in the middle of this range.

Upon learning of the remarkable calculations being performed at Princeton University, Wilson and Penzias immediately suspected that the mysterious source of static might actually be radiation from the cooled-off primordial fireball. Observations were made at a series of wavelengths and, as shown in Figure 17-2, the data fall directly along the theoretical curve for 2.7° above absolute zero. It is generally believed that this background radiation is the cooled-off "echo" of the big bang.

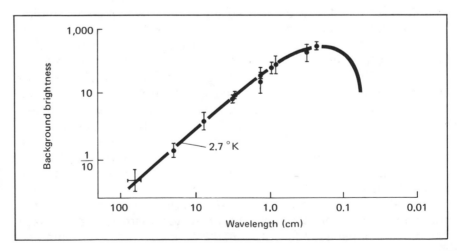

FIGURE 17-2. *The Background Radiation.* Weak radio waves coming from every part of the sky correspond to an overall temperature of the universe at 2.7° above absolute zero.

One of the most unusual properties of the 3° background is that it is incredibly *isotropic.* No matter what time of the day or night, no matter what time of the year, radio astronomers always detect exactly the same intensity of background radiation from all parts of the sky. This is puzzling to the astrophysicist because there is no apparent reason why the temperature in one direction in the universe should be exactly the same as in any other direction.

Various attempts to explain the isotropy of the 3° background radiation have met with various degrees of success. For example, in the late 1960s, Charles Misner at the University of Maryland proposed the *mixmaster universe* based on an unusual class of solutions to the Einstein field equations. The mixmaster universe undergoes violent gyrations in all directions and is supposed to describe the conditions shortly after the big bang. Although Misner's work gives important insight into the possible nature of the early universe, it does not fully account for the isotropy of the 3° background radiation. The essential important feature to realize is that since the background radiation is so smooth, something must have happened to smooth it out.

While the isotropy of the 3° background radiation stands in mute testimony to violent mixing and smoothing processes that must have occurred before the universe was one second old, the entire universe did *not* become completely homogeneous. If the entire universe had been completely homogenized 18 billion years ago, it would still be homogeneous today. But there are clumps of matter in the universe in the form of stars and galaxies, planets and people. Therefore the violent processes responsible for smoothing out the radiation field were not completely successful in smoothing out the matter in the universe.

In the early 1970s, the brilliant British astrophysicist Stephen W. Hawking began seriously thinking about the consequences surrounding the violent birth of the universe (see Figure 17-3). He first of all noted that under the usual processes of stellar evolution it is virtually impossible to create black holes with masses less than three solar masses. Dying stars with masses less than three suns become white dwarfs or neutron stars, as discussed in Chapter 7. The primary reason for this lower limit to the mass of an ordinary black hole is that its creation relies completely on the inward pull of gravity. Only if the mass of the dying star is greater than three suns can the sheer weight of trillions upon trillions of tons of matter pressing inward from all sides overpower all physical forces and produce a black hole. Anything containing less than three solar

FIGURE 17-3. *Hawking.* Stephen W. Hawking (left) discusses primordial black holes with the author at Caltech. (Courtesy of C. Caves)

masses cannot become a black hole simply because there are forces in nature, such as degenerate electron and neutron pressure, which can always halt the collapse.

This traditional view of the masses of black holes had been well understood since the mid-1960s. A brick or a watermelon could never become a black hole in the universe as we know it today. Such objects do not possess enough matter to create an overpowering gravitational field. But suppose a low-mass object could be *squeezed* down to a very small size. If a brick or watermelon could be squeezed down to an extremely small size, smaller than the size of an electron, it would disappear behind its own event horizon, thereby producing a very tiny black hole. While such processes are inconceivable today, Hawking realized that if the early universe was violent enough to smooth out the background radiation, conditions should have been violent enough to squeeze out many very tiny black holes! Hawking therefore postulated the existence of *primordial black holes* that could have masses much less than three suns. In fact, primordial black holes with masses as low as one hundred thousandth of a gram could have been created by the violent crushing events immediately following the big bang. It is therefore conceivable that numerous very tiny black holes are scattered around the universe.

By the mid-1970s, Hawking had come to realize that primordial black holes must have properties very different from ordinary large black holes. To see why this is so, consider an ordinary black hole formed from a massive dying star. The gravitational field of this black hole can be felt over millions of miles. Even very far from the black hole, planets could be orbiting the hole just as if they were going about an ordinary star. In the case of a primordial black hole, however, only a few feet above the hole space-time is almost perfectly flat. If you stand a few feet from a mountain consisting of a billion tons of rock, you do not feel yourself being pulled toward the mountain. Similarly, if you were to stand a few feet from a primordial black hole made from a billion tons of matter, you would hardly notice any gravitational attraction from the hole at all.

Regardless of the mass of a black hole, all black holes have a singularity surrounded by an event horizon. All black holes contain a place where there is infinite curvature of space-time. In the case of a massive black hole, such as those formed from dying stars, the curvature of space-time gradually increases as one approaches the singularity. However, in the case of a primordial black hole, the curvature of space-time is much more sudden. As shown in Figure 17-4, over a very short distance the curvature of space-time goes from being almost perfectly flat to being very highly warped. This means that the *tidal forces,* the forces which try to pull things apart, must be enormous very near a primordial black hole. Although the overall gravitational field is weak, the tidal stresses surrounding a tiny black hole are very much stronger than in the case of an ordinary black hole.

As noted near the end of Chapter 14, theoretical physicists often find it very useful to think of empty space as actually consisting of imaginary pairs of particles and antiparticles. Using this concept of virtual pairs, it is much easier to understand the processes of pair creation and annihilation which nuclear physicists observe in their laboratories.

Now consider empty space immediately surrounding a tiny, primordial black hole. As mentioned above, very powerful tidal forces are present in this region of space. But it is entirely reasonable to think of this empty space as actually consisting of virtual pairs of particles and antiparticles. Realizing this fact, Hawking discovered that the tidal forces are so intense that they can tear apart the virtual

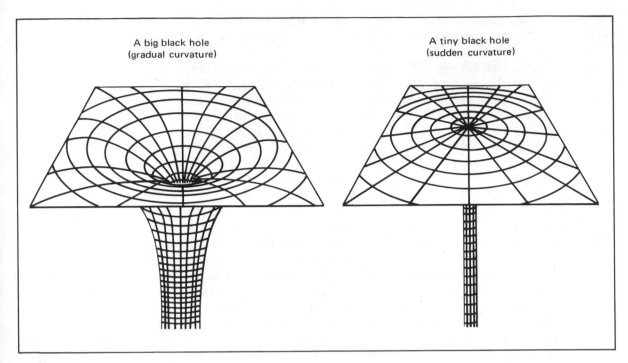

A big black hole
(gradual curvature)

A tiny black hole
(sudden curvature)

FIGURE 17-4. *Large and Small Black Holes.* The change in curvature of space-time near a primordial black hole is far more abrupt than in the case of an ordinary, massive black hole. The tidal forces surrounding a primordial black hole must be enormous.

pairs with such violence that the imaginary particles become real particles and actually appear in space. In other words, tiny black holes should emit great quantities of particles and antiparticles!

At this point you might guess that this process of creation of matter and antimatter by a black hole could lead to the emission of particles to the outside universe *only* if the process occurs above the hole's event horizon. In yet another major theoretical advance, in 1974, Hawking proved that this is not necessarily so. Even if the particles are created inside the hole, there is a probability that they can get through the event horizon to the outside universe. To understand why particles can *tunnel* through the event horizon, it is necessary that we should explore some basic ideas in *quantum mechanics.*

Back around the turn of the century when physicists began seriously thinking about atoms, electrons, protons and other subatomic particles, it was common to imagine that these particles were essentially little balls of matter. In the early, crude models of atoms it was quite acceptable to think of electrons, for example, as tiny billiard balls. This classical picture of subatomic particles came from nineteenth-century classical physics and is represented at the top of Figure 17-5.

As the study of atomic and nuclear physics progressed, it was realized that the billiard-ball picture of particles had inherent and fundamental limitations. For example, Werner Heisenberg discovered that when considering the very tiny distances involved inside atoms, it is impossible to know precisely where particles such as electrons really are. This fundamental inability to pinpoint the location of any subatomic particle was formulated in Heisenberg's *uncertainty principle*. The essential idea is that, at best, physicists can talk of the *probability* of a particle's being in a certain location at a certain time, but they can never be 100 percent sure, as they would be if the classical billiard-ball view were correct. As a result of these considerations, physicists found it much more useful and proper to think of subatomic particles as consisting of *wave packets* of matter, as represented at the bottom of Figure 17-5. This view of particles is one of the essential ideas in the field of physics called quantum mechanics.

When the quantum-mechanical approach to matter is taken, many phenomena can be understood that would be totally unexplainable with the old-fashioned billiard-ball picture. A good example involves the operation of transistors and diodes found in electronic equipment. Inside certain types of diodes, an electric field produces a *potential barrier* that is so strong that it would seem to prohibit electrons from getting from one side of the diode to the other. In this sense the potential barrier can be thought of as a "wall." With the billiard-ball picture, electrons would be expected simply to bounce off the wall, as shown at the top of Figure 17-6. On the other hand, if electrons are considered as wave packets of matter, there is a probability that the electrons can get through the potential barrier. This phenomenon is called *tunneling* and is shown at the bottom of Figure 17-6.

The gravitational field surrounding a black hole may be thought of as a potential barrier that classically prevents anything from getting out of the hole. In the case of a very massive black hole, the intense gravitational field extends over such a large distance that the potential barrier is very *thick*. The probability of a particle tunneling through such a potential barrier is virtually zero, as shown at the top of Figure 17-7. In a tiny, primordial black hole, however, the gravitational field is intense only over a very small distance. This means that the potential barrier surrounding a small black hole is very *thin* and, as shown at the bottom of Figure 17-7, there is a sizable proba-

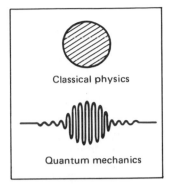

FIGURE 17-5. *A Subatomic Particle.* From the viewpoint of nineteenth-century classical physics, particles such as electrons or protons were thought of as being like tiny billiard balls. Quantum mechanics, developed during the twentieth century, describes subatomic particles in terms of wave packets.

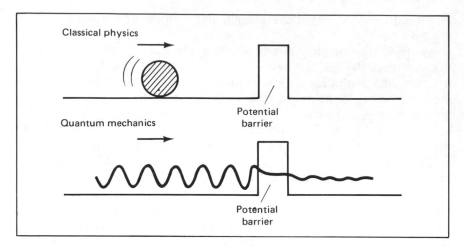

FIGURE 17-6. *Tunneling.* Classically, an electron should never be able to get through a strong potential barrier. Quantum mechanically, however, it is possible for subatomic particles to get from one side of the barrier to the other.

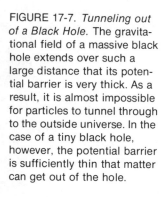

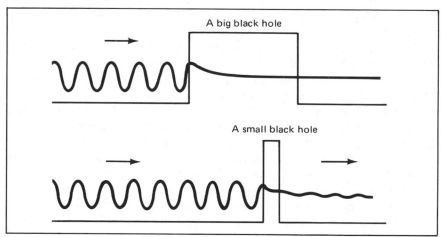

FIGURE 17-7. *Tunneling out of a Black Hole.* The gravitational field of a massive black hole extends over such a large distance that its potential barrier is very thick. As a result, it is almost impossible for particles to tunnel through to the outside universe. In the case of a tiny black hole, however, the potential barrier is sufficiently thin that matter can get out of the hole.

bility that particles can tunnel through to the outside universe. Particles and antiparticles created *inside* the event horizon can tunnel through the thin potential barrier of a small black hole and escape! This remarkable discovery that matter can escape from inside a black hole means that black holes behave like white holes. Following up this idea, Hawking in 1975 succeeded in proving that *small black holes are totally indistinguishable from small white holes!*

The fact that black holes emit matter and radiation means that it is possible to assign a *temperature* to a black hole. The temperature of the hole is a direct measure of the rate at which the hole emits

284

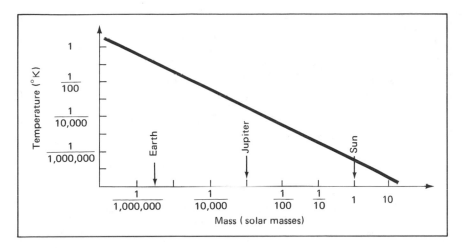

FIGURE 17-8. *Big Black Holes Are Cold.* Since it is almost impossible for particles to tunnel through the thick potential barriers surrounding large black holes, their temperatures are extremely low.

particles and radiation. Massive black holes have very thick potential barriers. The probability that anything can tunnel through such barriers is almost zero. Consequently, the temperature of massive black holes is almost zero. For example, the temperature of a black hole formed from a massive dying star is less than one millionth of a degree above absolute zero. Consequently, these quantum-mechanical effects discovered by Hawking are totally unimportant in massive black holes. Figure 17-8 shows the relationship between temperature and mass of large black holes. Black holes whose masses are greater than the mass of the earth have temperatures less than a tenth of a degree above absolute zero.

Big black holes are cold because they are surrounded by thick potential barriers that effectively prevent almost anything from getting out of the event horizon. Small black holes (if they exist) would have thin potential barriers. Particles and radiation can quantum mechanically escape from such holes and therefore their temperatures can be substantial. Figure 17-9 shows the relationship between the temperatures and masses of small black holes. As seen from the graph, a black hole whose mass is roughly the same as a typical asteroid's would have a temperature of about one hundred thousand degrees.

Very tiny black holes would have extremely thin potential barriers through which particles and radiation could easily escape to the outside universe. Consequently, the temperature of a very tiny black hole can be truly enormous. A black hole whose mass is a million tons emits so much matter and energy that its temperature is one

285

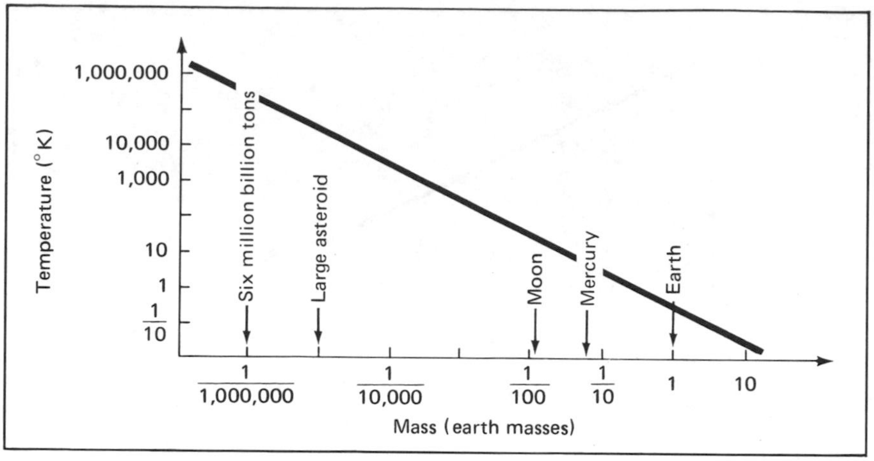

FIGURE 17-9. *Small Black Holes Are Hot.* Low-mass black holes are surrounded by thin potential barriers. The lower the mass, the thinner the barrier. Particles and radiation can tunnel through thin barriers and therefore small black holes can have substantial temperatures.

quadrillion degrees. A one-ton black hole has a temperature of one trillion billion degrees. Figure 17-10 shows the temperatures of black holes over a very wide range of masses.

When matter and radiation are emitted by a very small black hole, the hole must give up some of its mass. If a black hole emits 10 pounds of matter, its mass must decrease by exactly 10 pounds. This simple fact has a very important consequence. As a black hole emits matter and radiation, its mass decreases. As its mass decreases, the potential barrier becomes thinner, its temperature goes up, and the hole therefore radiates more particles and energy. The more it emits, the smaller it gets. The smaller it gets, the more it emits. This is a vicious cycle that completely eats up the black hole. Indeed, *black holes evaporate*. The process of evaporation escalates as the mass of the hole decreases. The escalation is so rapid for a very tiny black hole that during the final few seconds of evaporation the *black hole explodes*. The total amount of energy released during the final second of evaporation is equivalent to a *billion megaton hydrogen bombs!*

If the universe is 18 billion years old, and if primordial black holes were created during the big bang, then the very tiny black holes would have already evaporated by now. A very small black hole would have been so hot and would have emitted matter so rapidly that it simply could *not* exist for a very long time. It is therefore possible to speak of the *lifetime* of a black hole. A black hole made of 100 tons of matter is so hot that it could exist for only one ten-thousandth of a second before completely evaporating. A million-

286

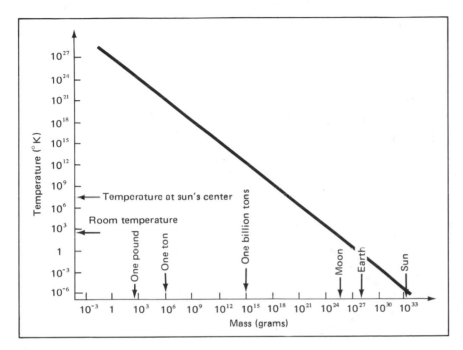

FIGURE 17-10. *The Temperatures of Black Holes.* The lower the mass of a black hole, the higher is its temperature. Very tiny black holes can have incredible temperatures.

ton black hole would take about three years to evaporate completely. A billion-ton black hole would last for about three billion years. Figure 17-11 shows the lifetimes of black holes of various masses.

The fact that the initial mass of a primordial black hole determines its lifetime means that the smallest black holes created during the big bang have already evaporated. Indeed, only those primordial black holes with masses greater than a few billion tons (10^{15} grams) could have survived up to the present time. Therefore, if scientists ever find primordial black holes in space, they would be at least as massive as a typical asteroid yet probably no bigger than an atom. These very tiny objects would be recognized because they emit incredible amounts of energy, probably in the form of very "hard" gamma rays.

The question of the possible discovery of primordial black holes in space is obviously directly related to how many such holes were probably created during the big bang. Late in 1975, Bernard Carr, who worked for several years with Hawking, presented strong arguments that primordial black holes could be rather plentiful. Specifically, Carr argues that it is possible that massive primordial black

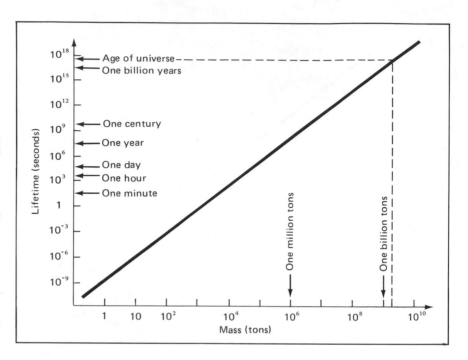

FIGURE 17-11. *The Lifetimes of Black Holes.* As a black hole emits particles and radiation, its mass decreases. As the mass decreases, the black hole emits more particles and radiation at a higher temperature. This escalating process means that all black holes eventually *evaporate.* During the final few microseconds of evaporation, a black hole emits an amount of energy equal to one billion megaton hydrogen bombs! Since the age of the universe is approximately 18 billion years, all black holes with masses less than a few billion tons should have evaporated by now.

holes (masses greater than a million suns) might exist at the centers of galaxies. If this is the case, then it is conceivable that a couple of tiny primordial black holes might be near or even *in* our solar system. It is indeed remotely possible that, unknown to us, a small primordial black hole could be in orbit about the sun.

If a primordial black hole were discovered in or near the solar system, the implications for us are tremendous. I believe that we now in principle possess the technology to go out and capture such a hole and bring it back to earth. With the hole placed in earth orbit, the energy from evaporation could be beamed down to the earth's surface in microwaves, thereby providing us with a vast amount of clean energy. The economic and political ramifications are staggering. Not only would we no longer be dependent on fossil fuels, but also a black hole orbiting the earth would render thermonuclear weaponry totally obsolete. It is therefore conceivable that the esoteric mathematical calculations of a few theoretical astrophysicists could dramatically influence the entire course of humanity.

Recent Theoretical Developments Concerning
Primordial Black Holes

(Note added in press)

Late in 1976, when this text was about to go to press, some important theoretical advances were made concerning the possible existence of numerous primordial black holes scattered throughout the universe. Fortunately, we were able to add this brief note.

In the early 1970s, Stephen Hawking proved that numerous tiny black holes could have been created during the big bang. By the mid-1970s, Dr. Hawking showed that, as a result of quantum mechanical effects, these primordial black holes should be emitting particles and radiation. The result of this emission of matter and energy is that primordial black holes *evaporate* and *explode.* During the explosion of a primordial black hole, the hole gives up all its mass in an incredibly brief and violent burst of highly energetic gamma-rays. As the discussion near the end of Chapter 17 mentions, all low-mass primordial black holes should have evaporated by now. Only the moderate-mass primordial black holes (that is, only those with masses greater than a few billion tons or 5×10^{14} grams) could have survived to the present time.

In 1975, George F. Chapline noted that the observed gamma-ray background could be used to estimate the maximum number of primordial black holes in the universe today. By assuming that *all* the observed gamma-ray background is caused by black hole evaporations, Don N. Page showed that there cannot be more than 300 million primordial black holes per cubic light year inside galaxies, provided the holes are clustered in galaxies. (This "upper limit" is reduced to only 300 primordial black holes per cubic light year if the holes are scattered uniformly throughout the universe.) Of course, much of the gamma-ray background may be due to effects other than the evaporation of primordial black holes, and therefore the number of these black holes existing today would be considerably less than the upper limits given above.

In October 1976, D. N. C. Lin, B. J. Carr, and S. M. Fall, who have been working on this matter, announced the results of their calculations; these show that the observed gamma-ray background does *not* necessarily limit the maximum number of primordial black holes that could conceivably exist today. Lin, Carr, and Fall proved that, under certain circumstances, very tiny primordial black holes created during the big bang could have grown so large that they would become moderate-mass primordial black holes (i.e., holes containing billions of tons of matter) before the universe was 1/10,000 second old. If this is the case, then enormous numbers of moderate-mass primordial black holes might be scattered throughout space. Because they have not yet completely evaporated and exploded, their existence remains undetected; they have not contributed to the gamma-ray background.

Tiny primordial black holes could have grown into moderate-mass holes *only* if the pressure during the earliest stages of the universe was truly enormous. Specifically, the growth of primordial black holes depends on the *equation of state* of the early universe. An equation of state simply relates pressure and density. If the pressure is so large that it equals the density, the equation of state is said to be "stiff." Normally, the pressure is less than the density and the equation of state is "soft." If circumstances immediately following the big bang were such that the equation of state of the universe was "stiff," then the pressures would have been so high (compared to the densities) that, due to pressure effects, matter would have been squeezed into tiny primordial black holes. These black holes would therefore have grown very rapidly into moderate-mass black holes.

After the universe was more than 1/10,000 second old, the equation of state must forever be "soft." From that time on, everything was so sufficiently dispersed that the equation of state could never be "stiff." Whether or not the equation of state of the universe was "stiff" *prior* to the first 1/10,000 second is very debatable. But Lin, Carr, and Fall have proved that *if* the equation of state of the universe was "stiff" for only 1 trillion-trillionth of a second (10^{-24} sec.) after the big bang, then pressures were sufficiently high for a long-enough period to cause *all* primordial black holes to grow to moderate masses (i.e., to have masses greater than a few billion tons). None of these primordial black holes would have been small enough to have completely evaporated by now. They would *all* be around today, some of them perhaps right here in our own solar system!

GLOSSARY

aberration (of starlight). The phenomenon whereby the positions of stars are displaced due to the motion of the observer.

absolute magnitude. A measure of the real brightness of a star; technically, the apparent magnitude a star would have at a distance of 10 parsecs.

absolute zero. A temperature of $0°K$ ($-273°C$); the lowest possible temperature.

acceleration. A change in velocity.

accretion. A process whereby an object gravitationally attracts material.

accretion disk. A disk of matter orbiting a black hole.

angstrom (Å). A very small unit of length; there are one hundred million angstroms in a centimeter.

angular momentum. A measure of the momentum associated with the rotation of an object.

antigravity. A gravitational field that repels objects and light rays.

antigravity universe. A universe in which gravity is repulsive; the universe on the "other" side of the ring singularity of a Kerr black hole.

aphelion. The point in the orbit of an object about the sun where the orbit is farthest from the sun.

apparent magnitude. A measure of the observed brightness of a star seen in the sky.

asteroid. A minor planet; one of many small objects orbiting the sun, usually at distances between the orbits of Mars and Jupiter.

astronomy. The science dealing with the nature and properties of the universe and its contents beyond the earth.

astrophysics. That part of astronomy dealing with the physics of the universe and its contents.

atom. The smallest particle of an element that retains the properties characterizing that element.

azimuthal angle. The angle measured from the axis of rotation.

barred-spiral galaxy. A spiral galaxy in which the spiral arms originate at the ends of a "bar" passing through the galaxy's nucleus.

big bang. The primordial explosion from which the universe presumably originated.

big bang cosmology. A cosmological model that states the universe began with a primordial explosion.

binary star. Two stars revolving about each other; a double star.

black hole. A highly warped region of space-time consisting of a singularity surrounded by an event horizon.

black hole bomb. A hypothetical explosive device involving the super-radiant scattering of light trapped around a Kerr black hole.

black hole evaporation. The process whereby a primordial black hole emits particles and radiation.

black hole explosion. The catastrophic result of the runaway evaporation of a primordial black hole.

blue sheet. A region of highly blueshifted light that has accumulated along an event horizon.

blueshift. A decrease in the wavelength of electromagnetic radiation.

causality (law of). The belief that effects must occur after their causes, not before.

Chandrasekhar limit. The upper limit to the mass of a white dwarf; approximately 1½ solar masses.

comet. A small object consisting of dust, gas, and ice, usually in a highly elliptical orbit about the sun.

conformal map. A drawing of space-time, made according to certain mathematical rules, that displays all regions of space-time.

conformal mapping. A technique of drawing space-time diagrams according to certain mathematical rules.

conic sections. A family of curves obtained by slicing a cone with planes at various angles.

coordinate system. A grid that can be used to denote positions in space or space-time.

coordinate time. Time measured by the clocks of an observer in flat space-time, far from any sources of gravity.

cosmic censorship. The idea that all singularities must be surrounded by an event horizon.

cosmic ray. Particles (usually protons or electrons) that strike the earth with exceedingly high speeds.

cosmology. A theory or study of the organization and evolution of the universe.

covariance (principle of). The belief that the most fundamental laws of physics can be stated in a way that is independent of the motion of an observer.

deferent. A circle in the Ptolemaic (geocentric) system along which epicycles move.

degenerate electron pressure. A pressure produced by electrons that resist further compression due to the Pauli exclusion principle.

degenerate gas. A gas in which all the possible states have been filled by the particles making up the gas.

degenerate neutron pressure. A pressure produced by neutrons that resist further compression due to the Pauli exclusion principle.

Doppler effect. The phenomenon whereby the wavelengths of radiation (or sound waves) are altered due to relative motion between the source and the observer.

Doppler shift. A shifting of spectral lines due to the Doppler effect.

double star. Two stars revolving about each other; a binary star.

dragging of inertial frames. The phenomenon whereby space-time is pulled around with a rotating body; the Lense-Thirring effect.

eclipse. The cutting off of all or part of the light of one body by another passing in front of it.

eclipsing binary. A double star whose orbit is oriented so that, as seen from earth, one star alternately passes in front of the other.

Einstein tensor. A tensor, constructed from the Ricci tensor, that appears in the field equations of general relativity.

electromagnetic field. A region of space containing both electric and magnetic fields.

electromagnetic field equations. A set of four equations that describe the behavior of electric and magnetic fields.

electromagnetic field tensor. A mathematical quantity expressed in four-dimensional space-time that fully describes an electromagnetic field.

electromagnetic radiation. Radiation consisting of electric and magnetic fields that move at the speed of light; these include radio, infrared, visible, ultraviolet, x-rays, and gamma rays.

electromagnetic spectrum. The entire array of electromagnetic radiation ranging from very short wavelength gamma rays to the longest wavelength radio waves.

electromagnetic theory. A branch of physics dealing with electric and magnetic phenomena.

electron. A negatively charged subatomic particle that normally moves about the nucleus of an atom.

ellipse. A conic section; the curve obtained by slicing all the way through a cone with a plane.

elliptical galaxy. An elliptically-shaped galaxy that does not exhibit any spiral structure.

"elsewhere." That region of space-time that is forever forbidden to material objects.

embedding diagram. A drawing of a spacelike hypersurface obtained by slicing through curved four-dimensional space-time.

epicycle. A circular orbit of a body in the Ptolemaic (geocentric) system.

293

equivalence principle. The idea that in a small region of space it is impossible to distinguish between gravitation and acceleration.

ergosphere. A region surrounding a Kerr black hole (between the static limit and the outer event horizon) in which it is impossible to be at rest.

"ether." A "substance," postulated by nineteenth-century physicists, that supports the wave motion of electromagnetic radiation.

event. The location in space-time of a phenomenon that occurs instantaneously at a particular position.

event horizon. A place in space-time where, according to a distant observer, time appears to stop; the surface of a black hole.

exit cone. An imaginary cone on the surface of a collapsing star that can be used to identify light rays capable of escaping from the star.

extreme Kerr hole. An uncharged, rotating black hole for which $M = a$.

extreme Reissner-Nordstrøm hole. An electrically charged nonrotating black hole for which $M = |Q|$.

field equations (in general relativity). A set of equations that describe the curvature of space-time by a gravitational field.

Fitzgerald contraction. The phenomenon whereby distances measured parallel to the direction of motion of a moving observer appear to have shrunk according to a stationary observer.

fusion (thermonuclear). The building up of heavier atomic nuclei from lighter ones.

future. That region of space-time into which material objects go.

future null infinity ($\mathscr{I}^+$). That region of space-time in the very distant future toward which all light rays go.

future timelike infinity (I^+). That region of space-time in the very distant future toward which the world lines of all material particles go.

galaxy. A huge collection of stars.

gamma rays. Photons of energy higher than those of x-rays; the most energetic form of electromagnetic radiation.

general theory of relativity. A theory of gravitation that expresses gravity in terms of the geometry of space-time.

geodesic. The shortest path between two events in curved space-time.

geodesic equations. A set of equations whose solution gives the shortest path (i.e., the geodesic) between two events in curved space-time.

gravitation. The phenomenon whereby material objects attract each other.

gravitational lens. The distortion of an image or the formation of multiple images by an intense gravitational field.

gravitational redshift. The slowing down of time as predicted by the general theory of relativity.

gravitational wave. Ripples in space-time that move with the speed of light.

gravitational wave antenna. A device designed to detect gravitational waves.

grey hole. A region of intense gravity consisting of a singularity and an event horizon out of which matter momentarily can emerge before plunging back in.

heliocentric. Centered about the sun.

helium burning. The thermonuclear conversion of helium into carbon.

Hertzsprung-Russell diagram. A plot of the luminosities of stars against their temperatures.

Hubble law. A relationship between the redshifts of galaxies and their distances.

hydrogen burning. The thermonuclear conversion of hydrogen into helium.

hyperbola. A conic section; a curve obtained by slicing a cone at a very steep angle.

hypersphere. A sphere in a hypothetical multi-dimensional "space."

hypersurface. A two-dimensional slice through four-dimensional space-time.

inner Lagrangian point. The point between the two Roche lobes in a binary.

interference fringes. Dark and light patterns produced when the waves in two beams of light either amplify or cancel out each other.

interval. The distance in space-time between two events.

invarient. A quantity that has the same value for all observers regardless of their state of motion.

irregular galaxy. An unsymmetrical galaxy.

isotropy. The quality of being the same in all directions.

Kepler's laws. Three laws, formulated by Johannes Kepler, that describe the motions of planets about the sun.

Kerr black hole. A black hole having mass and angular momentum; an uncharged, rotating black hole.

Kerr-Newman solution. A solution to the field equations of general relativity describing a charged, rotating black hole.

Kerr solution. A solution to the field equations of general relativity describing an uncharged, rotating black hole.

Kruskal-Szekeres diagram. A space-time diagram of a Schwarzschild black hole that displays the full geometry of the hole.

Lense-Thirring effect. A phenomena whereby space-time is "dragged" around a rotating body; the "dragging of inertial frames."

light. Electromagnetic radiation that is visible to the eye.

lightlike trip. A path in space-time that is inclined at 45° from the time axis; the path followed by a light ray.

light-year. The distance light travels in a vacuum in one year; roughly six trillion miles.

Lorentz transformations. A set of equations in special relativity that relate measurements of quantities, such as time, mass, and distance, made by observers who are moving relative to each other.

luxon. Something that always moves at a speed exactly equal to the speed of light (e.g., photons, neutrinos).

main sequence. A grouping of stars on the Hertzsprung-Russell diagram in which hydrogen burning is the primary source of energy.

main-sequence star. A star whose position on the Hertzsprung-Russell diagram places it on the main sequence.

mass. A measure of the total amount of material in a body.

massive black hole. A black hole whose mass is roughly between 100 and 1,000 suns.

mass-luminosity relation. A relationship between the masses and the luminosities of main-sequence stars.

mass-transfer. The process whereby matter is exchanged between stars in a binary.

mechanics. The branch of physics that deals with the behavior of material bodies.

Messier catalogue. A catalogue of nonstellar objects compiled by Charles Messier in 1787.

Michelson interferometer. A device invented by Albert A. Michelson originally intended to detect the motion of the earth through the ether.

Michelson-Morley experiment. An experiment first performed in 1888 by Albert A. Michelson and Edward W. Morley that proved the "ether" does not exist.

mixmaster universe. The theoretical model of a universe that undergoes violent gyrations in all directions.

molecule. A chemical combination of two or more atoms.

Mössbauer effect. A phenomena in nuclear physics whereby radioactive nuclei can be used as extremely accurate clocks.

naked singularity. A space-time singularity that is not surrounded by an event horizon.

nebula. A cloud of interstellar gas or dust.

negative space. That region of space on the "other" side of the ring singularity of a Kerr black hole.

neutron. A subatomic particle with no charge and with mass approximately equal to that of the proton.

neutron star. A collapsed star supported by degenerate neutron pressure.

Newton's laws. The laws of mechanics and gravitation formulated by Isaac Newton.

nova. A star that experiences a sudden outburst, temporarily increasing its luminosity by hundreds to thousands of times.

nuclear. Referring to the nucleus of the atom.

nucleus (of atom). The heavy part of an atom, which is composed of protons and neutrons, and about which the electrons revolve.

nucleus (of galaxy). The central concentration of stars, and possibly gas, at the center of a galaxy.

oblate spheroidal coordinates. A coordinate system well-suited for describing the geometrical properties of a Kerr black hole.

observer. Someone who measures quantities such as time intervals, distances, masses, velocities, etc.

orbit. The path of a body that is in revolution about another body or point.

pair creation. The phenomenon whereby pairs of particles and antiparticles are created.

parabola. A conic section; the curve obtained by slicing a cone with a plane parallel to one side of the cone.

parallax. The apparent shifting of an object due to the displacement of the observer.

parallax (stellar). The apparent shifting of a nearby star due to the motion of the earth around the sun.

parsec. A large unit of length equal to 3.26 light-years.

past. That region of space-time from which material objects come.

past null infinity ($\mathscr{I}^-$). That region of space-time in the very distant past from which all light rays come.

past timelike infinity (I^-). That region of space-time in the very distant past from which the world lines of all material particles originate.

Pauli exclusion principle. A law of physics that states that no two particles (such as electrons or neutrons) with the same spin and velocity can occupy the same place at the same time.

pendulum circular orbit. A type of circular light orbit in the negative space of a Kerr hole.

Penrose diagram. A conformal map that includes all of space-time.

Penrose mechanism. A method of extracting energy from a rotating black hole.

perihelion. The point in the orbit of an object revolving about the sun where the orbit is closest to the sun.

perturbation. A small deflection or change.

photon. A discrete unit of electromagnetic energy.

photon circle. A circular light orbit about a Schwarzschild black hole.

photon sphere. A sphere of circular light orbiting about a Schwarzschild black hole.

planetary nebula. A shell of gas that has been ejected by a star near the end of its life cycle.

positron. An antielectron; an electron with positive charge.

potential barrier. A region of space in which a force acts to inhibit the motions of particles.

297

primary cosmic rays. The cosmic-ray particles that arrive at the earth from beyond its atmosphere (as opposed to the secondary particles that are produced by collisions between primary cosmic rays and air molecules).

primordial black hole. A very small black hole (mass less than 1 sun) that could have been created during the big bang.

primordial fireball. The extremely hot gas presumed to comprise the entire universe immediately after the big bang.

proper length. The distance between two points as measured by an observer who is at rest with respect to the two points.

proper mass. The mass of an object as measured by an observer who is at rest with respect to the object.

proper time. Time measured by the clocks of a freely-falling observer in a gravitational field.

proton. A heavy subatomic particle that carries a positive charge, and one of the two principal constituents of the atomic nucleus.

protostar. A contracting sphere of gas in the process of becoming a star.

pulsar. A rapidly pulsating radio source.

quantum mechanics. That branch of physics dealing with the properties and the behavior of atoms and subatomic particles.

quasar. A star-like object believed to be at an enormous distance from our own galaxy; often a source of radio waves.

radio telescope. A telescope designed to detect and amplify radio waves from astronomical sources.

randomicity principle. The idea that a space-time singularity produces matter and energy in a totally random fashion.

red giant. A very large, cool star.

redshift. An increase in the wavelength of electromagnetic radiation.

Reissner-Nordstrøm black hole. A black hole having mass and charge; a nonrotating, charged black hole.

Reissner-Nordstrøm solution. A solution to the field equations of general relativity describing a nonrotating, charged black hole.

Ricci tensor. A tensor constructed from the Riemann curvature tensor that appears in the field equations of general relativity.

Riemann curvature tensor. A tensor containing detailed mathematical information about the geometry of a "space" of any number of dimensions.

ring singularity. The singularity of a Kerr black hole.

Roche lobes. Imaginary figure-eight surfaces surrounding the two stars in a binary denoting the effective limits of the stars' gravitational influence.

Schwarzschild black hole. A black hole containing mass only; a nonrotating, uncharged black hole.

Schwarzschild radius. The radius of the event horizon surrounding a Schwarzschild black hole.

Schwarzschild solution. A solution to the field equations of general relativity describing a nonrotating, uncharged black hole.

secondary cosmic rays. Secondary particles produced by interactions between primary cosmic rays from space and atomic nuclei in the earth's atmosphere.

singularity. A location of infinite space-time curvature; the center of a black hole.

solar system. The system of the sun and the planets, their satellites, the minor planets, comets, meteoroids, and other objects revolving around the sun.

solar wind. High-speed protons and electrons constantly being emitted from the sun.

spacelike infinity (I^0). That very distant region of space-time that can be reached only along spacelike trips.

spacelike trip. A path in space-time that is inclined at an angle greater than 45° from the time axis.

space-time. A four-dimensional continuum in which three dimensions are spatial and the fourth is temporal.

special theory of relativity. A covariant formulation of mechanics and electrodynamics in flat space-time.

spectral line. A thin dark line among the colors of a spectrum.

spectroscopic binary. A double star whose nature is inferred from the periodic shifting of spectral lines in its spectrum.

spectrum. The array of colors obtained by passing white light through a prism.

spiral galaxy. A flattened, rotating galaxy with pinwheel-like arms originating from the galaxy's nucleus.

star. A large, self-luminous sphere of gas.

static limit. A location (near a black hole) inside of which it is impossible to remain at rest.

stellar evolution. The changes that take place in the sizes, luminosities, structures, and so on, of stars as they age.

stellar model. The result of a theoretical calculation of physical conditions in a star.

stellar wind. High-speed protons and electrons constantly being emitted from a star.

stress-energy tensor. A tensor containing detailed information about matter and nongravitational fields that appears in the field equations of general relativity.

supermassive black hole. A black hole whose mass is greater than 100,000 suns.

supernova. A violent stellar explosion that can occur near the end of the life of a very massive star.

superradiant scattering. A phenomenon whereby light passing near a rotating black hole is amplified.

tachyon. A hypothetical object that always moves at a speed greater than the speed of light.

tardyon. An object that always moves at a speed less than the speed of light.

tensor. A mathematical quantity having certain specific mathematical properties.

tensor analysis. A branch of mathematics dealing with tensors.

thermonuclear energy. The energy released by thermonuclear reactions.

thermonuclear reaction. A reaction or a transformation that results from the high-speed collisions of nuclear particles.

time dilation. The phenomenon whereby clocks of a moving observer appear to run slow according to a stationary observer.

time machine. A hypothetical device that could be used to travel into the distant future or into the past.

timelike trip. A path in space-time that is inclined at an angle less than 45° from the time axis.

tunneling. A quantum-mechanical phenomenon whereby particles can penetrate a potential barrier.

twin paradox. A paradox that results in thinking about two observers moving relative to each other, each one saying that the other's clocks have slowed down.

Uhuru. The nickname for Explorer 42, an earth-orbiting astronomical satellite designed to detect x-rays.

uncertainty principle. The idea that it is impossible to know both the location and the speed of subatomic particles with infinite precision.

variable star. A star that varies in luminosity.

virtual pair. An imaginary particle-antiparticle pair that has not been materialized.

visual binary. A double star in which both members can be seen.

wave equation. An equation that describes the propagation of a wave.

wavelength. The distance between successive peaks (or valleys) in a wave.

wave packet. The quantum-mechanical picture of a particle in terms of a concentration of waves.

white dwarf. A very small, hot star near the final stage of its evolution.

white hole. A region of intense gravity consisting of a singularity and an event horizon out of which matter and energy emerge; the time-reversal of a black hole.

world line. The path of an object in space-time.

INDEX